Selected Papers from IEEE ICKII 2018

Selected Papers from IEEE ICKII 2018

Special Issue Editors

TeenHang Meen
Wenbing Zhao

MDPI • Basel • Beijing • Wuhan • Barcelona • Belgrade

MDPI

Special Issue Editors

TeenHang Meen
National Formosa University
Taiwan

Wenbing Zhao
Cleveland State University
USA

Editorial Office
MDPI
St. Alban-Anlage 66
4052 Basel, Switzerland

This is a reprint of articles from the Special Issue published online in the open access journal *Electronics* (ISSN 2079-9292) from 2018 to 2019 (available at: https://www.mdpi.com/journal/electronics/special_issues/ieee_ickii_2018)

For citation purposes, cite each article independently as indicated on the article page online and as indicated below:

LastName, A.A.; LastName, B.B.; LastName, C.C. Article Title. *Journal Name* **Year**, *Article Number*, Page Range.

ISBN 978-3-03921-273-6 (Pbk)
ISBN 978-3-03921-274-3 (PDF)

Contents

About the Special Issue Editors

TeenHang Meen was born in Tainan, Taiwan, on 1 August 1967. He received his B.Sc. degree in electrical engineering from National Cheng Kung University (NCKU), Tainan, Taiwan, in 1989 and his M.Sc. and Ph.D. degrees from Institute of Electrical Engineering, National Sun Yat-Sen University (NSYSU), Kaohsiung, Taiwan, in 1991 and 1994, respectively. He was the chairman of the Department of Electronic Engineering from 2005 to 2011 at National Formosa University, Yunlin, Taiwan. He received the excellent research award from National Formosa University in 2008 and 2014. Currently, he is a distinguished professor in the Department of Electronic Engineering, National Formosa University, Yunlin, Taiwan. He is also the president of International Institute of Knowledge Innovation and Invention (IIKII) and the chair of IEEE Tainan Section Sensors Council. He has published more than 100 SCI, SSCI, and EI papers in recent years.

Wenbing Zhao is a full professor of electrical engineering and computer science (EECS) at Cleveland State University (CSU), Cleveland, Ohio, USA. He obtained his B.Sc. and M.Sc. degrees in physics from Peking University, Beijing, China, in 1990 and 1993, respectively, and his M.Sc. and Ph.D. degrees in electrical and computer engineering from University of California, Santa Barbara, in 1998 and 2002, respectively. Prior to joining Cleveland State University in 2004, Dr. Zhao worked as a post-doctoral researcher at University of California, Santa Barbara, and as a senior research engineer/chief architect at Eternal Systems, Inc. (now dissolved), which he co-founded in 2000. Dr. Zhao has done research in several different areas, including fault tolerance computing, computer and network security, smart and connected healthcare, machine learning, Internet of Things, quantum optics, and superconducting physics. Currently, his research focuses on smart and connected healthcare. Dr. Zhao's recent research has been funded by the National Science Foundation, Ohio Bureau of Workers' Compensation, Ohio Department of Higher Education, Ohio Advancement Office (via the Ohio Third Frontier Program), US Department of Transportation (via CSU Transportation Center), Cleveland State University, and private companies.

electronics

MDPI

Editorial

Special Issue on Selected Papers from IEEE ICKII 2018

Teen-Hang Meen [1,]* and Wenbing Zhao [2]

[1] Department of Electronic Engineering, National Formosa University, Yunlin 632, Taiwan
[2] Department of Electrical Engineering and Computer Science, Cleveland State University, Cleveland, OH 44011, USA
* Correspondence: thmeen@nfu.edu.tw

Received: 2 July 2019; Accepted: 4 July 2019; Published: 5 July 2019

1. Introduction

Electronic Engineering and Design Innovations are both academic and practical engineering fields that involve systematic technological materialization through scientific principles and engineering designs. Technological innovation via Electronic Engineering includes electrical circuits and devices, computer science and engineering, communications and information processing, and electrical engineering communications. The Special Issue on "Selected Papers from IEEE ICKII 2018" was expected to select excellent papers presented at the International Conference on Knowledge Innovation and Invention 2018 (IEEE ICKII 2018) on the topics of electronics and its applications. This conference was held on Jeju Island, South Korea, 23–27 July, 2018, and it provided a unified communication platform for researchers from all of the world. The main goal of this Special Issue on "Selected Papers from IEEE ICKII 2018" is to discover new scientific knowledge relevant to the topic of electronics and its applications.

2. The Topic of Electronics and Its Applications

This special issue selected 5 excellent papers from 120 papers presented at IEEE ICKII 2018. The published papers are introduced as follows:

Chang et al. reported on "Robust Stable Control Design for AC Power Supply Applications" [1] and proposed an improved feedback algorithm by binary particle swarm optimization (BPSO)-based nonsingular terminal sliding mode control (NTSMC) for DC–AC converters. The NTSMC can create limited system state convergence time and allow singularity avoidance. The BPSO is capable of finding the global best solution in real-world applications, thus optimizing NTSMC parameters during digital implementation. The association of NTSMC and BPSO extends the design of the classical terminal sliding mode to converge to non-singular points more quickly and introduce optimal methodology to avoid falling into local extremum and low convergence precision. Simulation results show that the improved technique can achieve low total harmonic distortion (THD) and fast transients with both plant parameter variations and sudden step load changes. Experimental results of a DC–AC converter prototype controlled by an algorithm based on digital signal processing have been shown to confirm mathematical analysis and enhanced performance under transient and steady-state load conditions. Since the improved DC–AC converter system has significant advantages in tracking accuracy and solution quality over classical terminal sliding mode DC–AC converter systems, this paper will be applicable to designers of relevant robust control and optimal control techniques.

Jiang et al. reported on "A Mixed Deep Recurrent Neural Network for MEMS Gyroscope Noise Suppressing" [2] to improve the navigation accuracy of inertial navigation systems, where one effective approach is to model the raw signal noise and suppress it. Commonly, an inertial measurement unit is composed of three gyroscopes and three accelerometers, and among them, the gyroscopes play an important role in the accuracy of the inertial navigation system's navigation solutions. Motivated by this problem, in this paper, an advanced deep recurrent neural network was employed and evaluated in noise modeling of a micromechanics system gyroscope. Specifically, a deep long short-term memory recurrent neural network and a deep-gated recurrent unit–recurrent neural network were combined together to construct a two-layer recurrent neural network for noise modeling. In this method, the gyroscope data were treated as a time series, and a real dataset from a micromechanics system inertial measurement unit was employed in the experiments. The results showed that, compared to the two-layer long short-term memory, the three-axis attitude errors of the mixed long short-term memory–gated recurrent unit decreased by 7.8%, 20.0%, and 5.1%. When compared with the two-layer gated recurrent unit, the proposed method showed 15.9%, 14.3%, and 10.5% improvement. These results supported a positive conclusion on the performance of the designed method, specifically, the mixed deep recurrent neural networks outperformed the two-layer gated recurrent unit and the two-layer long short-term memory recurrent neural networks.

Wang et al. reported on "Operational Improvement of Interior Permanent Magnet Synchronous Motor Using Fuzzy Field-Weakening Control" [3] and proposed that the fuzzy control design of maximum torque per ampere (MTPA) and maximum torque per voltage (MTPV) for the interior permanent magnet synchronous motor (IPMSM) control system is capable of reducing computation burden, improving torque output, and widening the speed range. In the entire motor speed range, three control methods, i.e., the MTPA, flux weakening, and MTPV methods may be applied depending on current and voltage statuses. The simulation using MATLAB/Simulink is first conducted and then in order to speed up the development, hardware-in-the-loop (HIL) is adopted to verify the effectiveness of the proposed fuzzy MTPA and MTPV control for the IPMSM system.

Zhang et al. reported on "Hardware Implementation for an Improved Full-Pixel Search Algorithm Based on Normalized Cross Correlation Method" [4] to propose an improved full pixel search algorithm based on the normalized cross correlation (NCC) method considering hardware implementation. According to the field programmable gate array (FPGA) simulation results, the speed of hardware design proposed in this paper is 2000 times faster than that of software in single-point matching, and 600 times faster than software in multi-point matching. The speed of the presented algorithm shows an increasing trend with the increase of the template size when performing multipoint matching.

Lee et al. reported on "Design and Realization of a Compact High-Frequency Band-Pass Filter with Low Insertion Loss Based on a Combination of a Circular-Shaped Spiral Inductor, Spiral Capacitor and Interdigital Capacitor" [5] to propose bandpass filter (BPF) is connecting an interdigital and a spiral capacitor in series between the two symmetrical halves of a circular intertwined spiral inductor. For the mass production of devices, and to achieve a higher accuracy and a better performance compared with other passive technologies, we used integrated passive device (IPD) technology. IPD has been widely used to realize compact BPFs and achieve the abovementioned. The center frequency of the proposed BPF is 1.96 GHz, and the return loss, insertion loss, and transmission zero are 26.77 dB, 0.27 dB, and 38.12·dB, respectively. The overall dimensions of BPFs manufactured using IPD technology are $984 \times 800 \ \mu m^2$, which is advantageous for miniaturization and integration.

Acknowledgments: The guest editors would like to thank the authors for their contributions to this Special Issue and all the reviewers for their constructive reviews. We are also grateful to Michelle Zhou, the Managing Editor of Electronics, for her time and efforts on the publication of this Special Issue for Electronics.

Conflicts of Interest: The authors declare no conflict of interest.

References

1. Chang, E.-H.; Yang, S.-C.; Wu, R.-C. Robust Stable Control Design for AC Power Supply Applications. *Electronics* **2019**, *8*, 419. [CrossRef]
2. Jiang, C.; Chen, Y.; Chen, S.; Bo, Y.; Li, W.; Tian, W.; Guo, J. A Mixed Deep Recurrent Neural Network for MEMS Gyroscope Noise Suppressing. *Electronics* **2019**, *8*, 181. [CrossRef]
3. Wang, M.-S.; Hsieh, M.-F.; Lin, H.-Y. Operational Improvement of Interior Permanent Magnet Synchronous Motor Using Fuzzy Field-Weakening Control. *Electronics* **2018**, *7*, 452. [CrossRef]

4. Zhang, G.; Kuang, Z.; Wei, S.; Huang, K.; Liang, F.; Yang, C.-F. Hardware Implementation for an Improved Full-Pixel Search Algorithm Based on Normalized Cross Correlation Method. *Electronics* **2018**, *7*, 428. [CrossRef]

5. Lee, K.-H.; Kim, E.-S.; Liang, J.-G.; Kim, N.-Y. Design and Realization of a Compact High-Frequency Band-Pass Filter with Low Insertion Loss Based on a Combination of a Circular-Shaped Spiral Inductor, Spiral Capacitor and Interdigital Capacitor. *Electronics* **2018**, *7*, 195. [CrossRef]

electronics

MDPI

Article

Robust Stable Control Design for AC Power Supply Applications

En-Chih Chang, Sung-Chi Yang and Rong-Ching Wu *

Department of Electrical Engineering, I-Shou University, No.1, Sec. 1, Syuecheng Rd., Dashu District, Kaohsiung City 84001, Taiwan; enchihchang@isu.edu.tw (E.-C.C.); r020313@gmail.com (S.-C.Y.)
* Correspondence: rcwu@isu.edu.tw; Tel.: +886-7-6577711 (ext. 6636); Fax: +886-7-6577205

Received: 3 December 2018; Accepted: 9 April 2019; Published: 10 April 2019

Abstract: This paper applies modified feedback technology to carry out the exact steady-state and fast transient in a high-performance alternating current (AC) power supply. The presented scheme displays the virtues of a finite-time convergence control (FTCC) and a discrete grey prediction model (DGPM). The FTCC, derived from a terminal sliding-mode (TSM) design principle, can produce the finite system-state convergence time and evade the singularity. It is noteworthy that the chattering/steady-state error around the FTCC may occur because of the overestimated or underestimated uncertainty bound. The DGPM with the bound estimate ability is integrated into the FTCC to cope with internal parameter variations and external load disturbances. The less chattering and steady-state error can be obtained, providing more robust performance in the AC power supply. The combination of the FTCC and the DGPM extends the standard TSM design for the purpose of faster singularity-free convergence, as well as introducing the grey modeling method in the case of a more exact uncertainty estimate. The modified control technology has a high-precision tracking performance and a fast convergent speed. Simulated and experimental results point out that the modified control technology can effectuate low total harmonic distortion (THD) and fast dynamic response in the presence of rectifier loads and abrupt step load changes.

Keywords: finite-time convergence control (FTCC); discrete grey prediction model (DGPM); chattering; AC power supply; total harmonic distortion (THD)

---

1. Introduction

Alternating current (AC) power supplies have been employed as an important unit for power conversion systems, such as uninterruptible power systems, solar photovoltaic systems, and wind turbine generating systems [1,2]. A high-performance AC power supply should comprise the following: (1) low harmonic distortion for linear/nonlinear loads. The IEEE standard 519-1992 suggests that the voltage total harmonic distortion (THD) is below 5%. (2) As applied to abrupt load changes, the dynamic response with smaller voltage sag and faster transient recovery time. According to IEEE standard 1159-1995, the voltage sag specifies a decrease in RMS (root mean square) voltage/current at the power frequency for durations from 0.5 cycles to one minute. The values of normal voltage sags are between 0.1 and. 0.9 per unit. (3) Nearly zero steady-state tracking errors. Sliding mode control (SMC) can provide the insensitivity to system uncertainties [3,4]; a number of SMCs presented for the AC power supply have been performed [5–7]. A fixed switching frequency sliding mode is applied to uninterruptible power supplies. The standard SMC is employed and the distorted output-voltage yields under nonlinear loading [5]. To enhance the system performance of the AC power supply, a standard SMC with an adaptive method is designed. Although the good performance in the steady-state and the transient can be obtained, the presented algorithm is complicated [6]. The combination of the standard SMC and the sensor number reduction is developed for three-phase inverters. The sophisticated hardware design is improved, but the chattering phenomenon still exists [7]. As previously mentioned, these standard SMC approaches display linear sliding surfaces, leading to a non-finite-time convergence. To raise the convergence speed, a finite-time convergence control (FTCC) is used with the nonlinear sliding surface in this paper. By appropriately designing the FTTC parameters, the system states will reach the sliding surface and converge to the equilibrium within a finite time, yielding a stable closed-loop system [8–10]. Nevertheless, the chattering will appear in the AC power supply output if a highly nonlinear load is applied. The chattering brings a high harmonic distortion and thermal breakdown in transistors, thus leading to instability and unreliability of the AC power supply system. Several approaches, such as observer methodology and adaptive control, for removing the chattering have strived to estimate the bounds of system uncertainties. These approaches indeed reduce the chattering, but the control designs incur long tracking times [11–13]. Since the 1980s, the discrete grey prediction model (DGPM) has tempted broad research interests because of its efficient and fast computation [14,15]. The DGPM just needs a few sampled data to depict the tendency of the time-series data from the anterior system dynamics, acquiring the dependable and acceptable forecast accuracy [16–19]. The DGPM is thus employed to lessen the chattering when the dynamic system with uncertainty bounds is overestimated. Using the modified control technology, the tracking errors can be minimized and the AC power supply creates the low distorted harmonics, the fast dynamics, the chattering reduction, and the steady-state error mitigation. Though the eventual performance results of the modified control system do not top the THD results of recent former research, the reinforcement of the FTCC methodology has produced a robust and more exact estimate of the uncertainty bounds. As can be noticed, the presented association of the FTCC and the DGPM contributes to a closed-loop feedback AC power supply with the small steady-state distortion and fast response under different load cases. The competence of the modified control technology is ratified via a digital implementation on a digital signal processing (DSP)-based AC power supply and the modified control system is also simulated using MATLAB/SIMULINK software.

2. System Modeling

Figure 1 illustrates the system block diagram of the AC power supply, which comprises transistor switches, an LC (inductor capacitor) filter, and a load. The v_o denotes the output-voltage, v_d stands for the desired sine wave, $e_1 = v_o - v_d$ is the voltage error, and R represents the load. Using a high switching frequency, the AC power supply and its PWM (pulse width modulation) are frequently modeled as a constant gain K_{pwm}. Therefore, the error dynamics can be formulated as

$$\begin{cases} \dot{e}_1 = e_2 \\ \dot{e}_2 = -a_1 e_1 - a_2 e_2 + bu - H \end{cases}, \tag{1}$$

where a_1 represents $1/LC$, a_2 stands for $1/RC$, b is K_{pwm}/LC, and $H = v_d/LC + \dot{v}_d/RC + \ddot{v}_d$ signifies the system uncertainty. It is noteworthy that the component values for the LC filter of the AC power supply can be chosen by the suggested methods as follows [20–22]. (i) Choose the switching frequency [21,22]—to decrease the size of the filter, a high enough switching frequency between 3 kHz to 15 kHz is frequently selected for IGBT (insulated gate bipolar transistor) switches. (ii) Choose a factor related to the cut-off frequency of the LC filter [20]—if the factor is lager, there is a great degradation and a small magnification in the switching frequency and the fundamental frequency, respectively. The minimum of the factor can be computed while the modulation value is recommended below 0.95. (iii) Choose a factor related to both switching frequency and inductor ripple current [22]—the inductor ripple current from 20% to 40% is an advisable range. Then, from (8), (20), (25), and (26) in the work of [20], the factor can be selected, and L and C component values are computed.

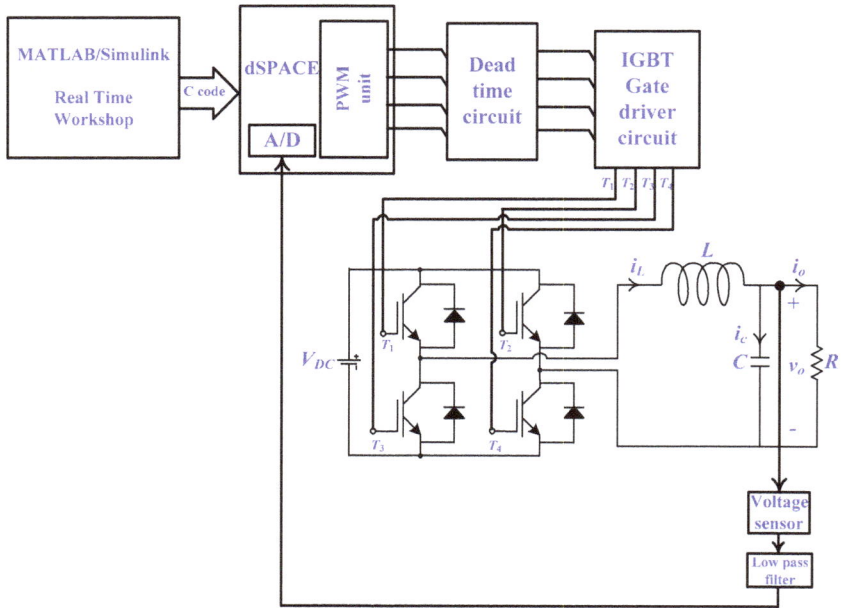

Figure 1. A control structure of an alternating current (AC) power supply system. PWM—pulse width modulation; IGBT—insulated gate bipolar transistor.

The control signal u in (1) has to be designed adequately, so as to coerce e_1 and e_2 to zero. Namely, the FTCC impels the system tracking behavior to converge to the origin in a finite time. Nevertheless, the load condition of the AC power supply may be a sudden large step change or a severe nonlinearity, the FTCC system effortlessly has the chattering or steady-state error, incurring in an aberrant tracking. As described in the introduction section, considerable studies have been conducted to reveal a diminution in the amount of the chattering or to make the steady-state error as small as possible. Recently, the practical application of the prediction methods has swiftly become a hot topic in both engineering and science. On the basis of such a prompting, it will be a good notion to introduce predictive modeling techniques into the FTCC design, affording better system robustness and an alternative reference to researchers interested in AC power supply applications. The FTCC with the DGPM assistance is presented to enhance the classic FTCC for the chattering/steady-state error mitigation providing a more exact tracking. The AC power supply system using this modified control technology allows a higher performance AC output-voltage in response to uncertain perturbations.

3. Control Technology Design

For the tracking error dynamics (1), the terminal sliding function is written as

$$s = \dot{e}_1 + \delta \cdot e_1^{\lambda}, \tag{2}$$

where $\delta > 0$ and $0 < \lambda < 1$.

The $s = 0$ and e_1 are reached within a finite time. The control law u can be designed to insure the subsistence of the TSM as follows:

$$u = u_{equ} + u_{ft}, \tag{3}$$

with the items of u_{equ} and u_{ft} as follows:

$$u_{equ} = b^{-1}[a_1 e_1 + a_2 e_2 - \delta \cdot (\lambda e_1^{\lambda-1} \cdot e_2)], \tag{4}$$

$$u_{ft} = -b^{-1}[\Omega sign(s)], \; \Omega > |H|, \tag{5}$$

where the u_{equ} is the equivalent control constituent and supervises the unperturbed dynamics, thus letting $s = 0$ and $\dot{s} = 0$. u_{ft} stands for the sliding control constituent with the disturbance rejection and, in consequence, the state behavior can arrive at the sliding mode $s = 0$ and accomplish a finite system-state convergence time. However, there are the following problems occurring in the equivalent control constituent. (1) u_{equ} containing the $e_1^{\lambda-1} e_2$ may induce a singularity if $e_2 \neq 0$, while $e_1 = 0$ and $0 < \lambda < 1$. The singularity leads to an unbounded control signal and the stability of the feedback system. (2) As a matter of fact, $e_1^{\lambda-1}$ may generate an imaginary number under the constraint $0 < \lambda < 1$.

To subdue the singularity problem, the following FTCC is formed as

$$s = e_1 + \frac{e_2^\rho}{\xi}, \; \xi > 0, \; 1 < \rho < 2. \tag{6}$$

Then, a sliding-mode reaching equation $\dot{s} = -\varepsilon_1 s - \varepsilon_2 |s|^\alpha sign(s)$ is utilized; thereupon, the control law u is restated as

$$u = u_{equ}^{new} + u_{ft}^{new}, \tag{7}$$

with

$$u_{equ}^{new} = b^{-1}[a_1 e_1 + a_2 e_2 - \frac{\rho}{\xi} \cdot e_2^{2-\rho}], \tag{8}$$

$$u_{ft}^{new} = -b^{-1}[\varepsilon_1 s + \varepsilon_2 |s|^\alpha sign(s)], \varepsilon_1, \varepsilon_2 > 0, 0 < \alpha < 1, \tag{9}$$

where u_{equ}^{new} represents the equivalent control without the singularity that conducts the system dynamics. u_{ft}^{new} represents the sliding control with the fast convergence, and can arrest the influence of system uncertainties. The control law u expressed in (7) is modified by the annexation of the DGPM (u_{dgp}), which can relieve the chattering in the AC power supply system. The modeling operation of the DGPM is illustrated in the following:

Step 1: Input the prime data

The prime data sequence is presumed as

$$x_p^{(0)} = \left\{ x_p^{(0)}(k), \; k = 1, \, 2, \, n \right\}, \tag{10}$$

where n is the number of the recorded data.

Step 2: Applying the mapping generating operation (MGO)

The employment of the MGO can map the prime data sequence $x_p^{(0)}$ onto the non-negative sequence $x_{mg}^{(0)}$ because of the positive or negative data sequence existing in the control system:

$$x_{mg}^{(0)} = \left\{ x_{mg}^{(0)}(k), \; k = 1, \, 2, \, n \right\}. \tag{11}$$

The relatedness $x_p^{(0)}$ and $x_{mg}^{(0)}$ can be described as

$$x_{mg}^{(0)} = MGO(x_p^{(0)}(k)) = \beta + \delta x_p^{(0)}(k), \beta, \delta > 0. \tag{12}$$

Step 3: Applying the accumulated generating operation (AGO)

The first-order AGO sequence can be acquired by using the AGO on $x_{mg}^{(0)}$ as follows:

$$x_{mg}^{(1)} = \text{AGO}(x_{mg}^{(0)}(k)) = \sum_{i=1}^{j} x_{mg}^{(0)}(i) , j = 1, 2, \cdots, n. \tag{13}$$

Step 4: Grey model

By employing the accumulated data sequence, $x_{mg}^{(1)}$, a first-order ordinary differential grey model is established as

$$\frac{d}{dt} x_{mg}^{(1)}(t) + a_{mg} x_{mg}^{(1)}(t) = b_{mg}, \tag{14}$$

where a_{mg} indicates the developing coefficient and b_{mg} is the grey input.

In order to attain the grey background value, the data sequence is formulated by employing the following MEAN generating operation to the $x_{mg}^{(1)}$.

$$\begin{aligned} z_{mg}^{(1)}(k) &= \text{MEAN}(x_{mg}^{(1)}) \\ &= 0.5 \cdot (x_{mg}^{(1)}(k) + x_{mg}^{(1)}(k-1)) , k = 2, 3, \cdots, n \end{aligned} \tag{15}$$

While the sampling interval is one unit, the differential of the generating sequence $x_{mg}^{(1)}$ can be expressed as

$$x_{mg}^{(0)}(k) + a_{mg} z_{mg}^{(1)}(k) = b_{mg} , k = 2, 3, \cdots, n. \tag{16}$$

In order to decide the values of a_{mg} and b_{mg}, (16) can be written as

$$Y = B\psi. \tag{17}$$

where $\psi = \begin{bmatrix} a_{mg} \\ b_{mg} \end{bmatrix}$, $B = \begin{bmatrix} -z_{mg}^{(1)}(2) & 1 \\ -z_{mg}^{(1)}(3) & 1 \\ \vdots & \vdots \\ -z_{mg}^{(1)}(n) & 1 \end{bmatrix}$, and $Y = \begin{bmatrix} x_{mg}^{(0)}(2) \\ x_{mg}^{(0)}(3) \\ \vdots \\ x_{mg}^{(0)}(n) \end{bmatrix}$.

By the least-squares method, the estimated parameters a_{mg} and b_{mg} can be solved as

$$\psi = \begin{bmatrix} a_{mg} \\ b_{mg} \end{bmatrix} = (B^T B)^{-1} B^T Y. \tag{18}$$

Substituting (18) into the differential equation, the count of the forecasted value yields

$$\hat{x}_{mg}^{(1)}(k+1) = \frac{b_{mg}}{a_{mg}} + (x_{mg}^{(0)}(1) - \frac{b_{mg}}{a_{mg}})e^{-a_{mg}(k)}. \tag{19}$$

Step 5: Applying the inverse accumulated generating operation (IAGO)

The data sequence $\hat{x}_{mg}^{(0)}(k+1)$ can be obtained by applying the IAGO on the $\hat{x}_{mg}^{(1)}(k+1)$ below.

$$\begin{aligned} \hat{x}_{mg}^{(0)}(k+1) &= \text{IAGO}(x_{mg}^{(1)}(k)), k = 2, 3, \cdots, n \\ &= \hat{x}_{mg}^{(1)}(k+1) - \hat{x}_{mg}^{(1)}(k) \\ &= (1 - e^{a_{mg}}) \cdot (x_{mg}^{(0)}(1) - \frac{b_{mg}}{a_{mg}}) \cdot e^{-a_{mg}(k)} \end{aligned} \tag{20}$$

Step 6: Applying the inverse mapping generating operation (IMGO)

By using the IMGO, the forecasted value of the prime data sequence $\hat{x}_p^{(0)}$ can be stated as

$$\hat{x}_p^{(0)}(k+1) = (1 - e^{a_{mg}}) \cdot (x_p^{(0)}(1) - \frac{b_{mg}}{a_{mg}})e^{-a_{mg}(k)} - \beta, \tag{21}$$

where β can eschew a negative sequence $x_{mg}^{(0)}(k)$. Thus, the control law of (13) is re-described as

$$u(k) = u_{equ}^{new}(k) + u_{ft}^{new}(k) + u_{dgp}(k), \tag{22}$$

where the annexed compensation part, that is, discrete grey prediction control, u_{dgp}, is capable mitigating the phenomenon of the chattering.

$$u_{dgp}(k) = \begin{cases} 0 & , |\hat{s}(k)| < \Delta \\ \Xi \hat{s}(k)sign(s(k)\hat{s}(k)) & , |\hat{s}(k)| \geq \Delta \end{cases}, \tag{23}$$

where Ξ indicates a constant, $\hat{s}(k)$ connotes the forecasted value of $s(k)$, and Δ is the system boundary.

4. Simulation and Experimental Results

The performance of the presented AC power supply is investigated through the simulations and experiments. The system parameters of the AC power supply are offered in Table 1. Figures 2 and 3 plot the simulated waveforms under a step change in load (from 12 ohm to no load) obtained using the modified control technology and the classic FTCC, respectively. Expectably, the modified control technology achieves a fast and an exact trajectory tracking, that is, the swift transient response of the output-voltage can be acquired and there is approximately no output-voltage swell at the firing angle. Reversely, the classic finite-time convergence controlled AC power supply system reveals an observable output-voltage swell at the firing angle. To examine a stricter situation (abrupt load change from no load to 12 ohm), the simulated waveforms obtained using the modified technology and the classic FTCC are shown in Figures 4 and 5, respectively. Contrary to the classic FTCC, the modified control technology displays a slight voltage slump and a swift output-voltage recovery, thus corroborating the finite-time reachable sliding surface. Because of the effectual compensation of the DGPM, after a tiny instant voltage slump (3 V$_{rms}$), the output voltage with the modified control technology can be reverted to the sinusoidal reference voltage; nevertheless, the classic FTCC brings a great voltage slump (32 V$_{rms}$). Figure 6 depicts that the modified control technology is capable of enduring the random variations of the filter parameters L and C from 20% to 150% and 20% to 150%, respectively, of the nominal values under a resistive load of 12 ohm. The classic FTCC shown in Figure 7 generates the sensitivity with conspicuous wobble to the LC variation, which degrades the system's robustness. The simulated comparison of the voltage slump and the %THD under a step loading and an LC variation is offered in Table 2. Figure 8 shows the different load situations used in the experiments, in order to test the transient and steady-state behavior of the AC power supply. Figures 9 and 10 illustrate the experimental waveforms obtained using the modified control technology and the classic FTCC in the presence of step load changes (from no load to full load), respectively. Figure 9 exhibits a petty voltage slump with a swift retrieval time, but the classic FTCC system revealed in Figure 10 has a poor transient and the recovery of the voltage slump takes a long time. The output voltage of the modified control system can reach 110 V RMS sine reference in the back of the 5 V$_{rms}$ puny voltage slump. Nevertheless, there is a nearly 30 V$_{rms}$ output voltage slump in the classic FTCC system, producing a disappointing transient response. When a full resistive load of 12 ohm is applied, the system performance in response to the random variations (20%~150% of the nominal value) of the filter parameters is also explored. The experimental output voltage shown in Figure 11 obtained using the modified control technology gives the tolerable ability of the larger parametric variations. For the classic FTCC, from the start to the termination of the waveform, an output voltage distortion shown in Figure 12 is seen with the sensitivity. The experimental result of the modified controlled AC power supply under a rectifier load (a 270 μF capacitor and a 35 Ω resistor in parallel) is displayed in Figure 13. Though the load current shows high spikes, the distortion of the output-voltage is tiny (voltage %THD = 1.35%). Inversely, the results yielded by the classic finite-time convergence controlled AC power supply under the identical load situation are illustrated in the Figure 14 with a severely distorted waveform (voltage %THD = 8.92%). Table 3 compares the experimental dissimilarity between the modified control technology and the classic FTCC. The experimental tracking errors of the modified control technology, traditional (standard) SMC, and improved SMC are provided in Figure 15. The modified controlled system creates a swift convergence to the origin in a short time and is nearly free of oscillation, as compared with the traditional SMC and the improved SMC. Really, a superior tracking preciseness, a minor harmonic distortion, and a quicker convergence speed were obtained by the modified control technology, which is suitable for the use of the AC power supply. In addition, the classic FTCC law in (3) contains the equivalent control term with the occurrence of the singularity, and the sliding control term with the phenomenon of the chattering. The singularity can be solved by using the modified control technology shown in (8); nevertheless, the parameter Ω that exists in (5) is frequently more than or equal to the interference $|H|$. The sliding control term of the classic FTCC is replaced by (9) and (23). Under the circumstance of the unknown upper bound of the parameter uncertainties and the external disturbances, the feedback gain of the over-conservatism may generate a large chattering in the classic FTCC. To reduce the gain magnitude and the chattering that exists in the sliding control term, the compensation of (23) can be aroused once $|s|$ overruns the boundary layer width. The feedback gain shown in (9) of the over-conservatism in the modified control technology is divided into two terms, containing the sliding control term and DGPM compensator. The gain magnitude of the modified control technology will become smaller than that of the classic FTCC if the upper bound of the parameter uncertainties and the external disturbances is unknown. Thereby, a robust system performance with reduced chattering in the face of high system uncertainties can be yielded by using the modified control technology. A brief summary of the simulated and experimental results is given to expound the dissimilitude between the modified control technology and the classic FTCC. In practice, the classic FTCC switching gain is proportional to the perturbation upper bound/system uncertainty. The greater switching gain endeavors to stabilize the classic FTCC system if the stern parametric uncertainties and external perturbations transpire. This may bring an undesirable acute chattering. So, the DGPM attempts to adjust the switching gain well, displaying moderation of the chattering and reinforcement of the system performance.

Table 1. System parameters. DC—direct current.

DC-Link Voltage	$V_{DC} = 200$ V
Filter inductor	$L = 0.5$ mH
Filter capacitor	$C = 20$ μF
Resistive load	$R = 12$ Ω
Output voltage and frequency	$v_o = 110$ V_{rms}, $f = 60$ Hz
Switching frequency	$f_s = 15$ kHz

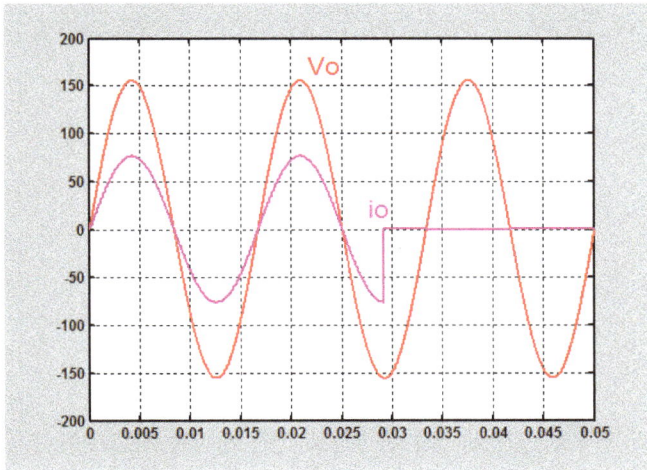

Figure 2. Simulated waveforms under the load suddenly turn off for the modified control technology (50 V/div; 10 A/div).

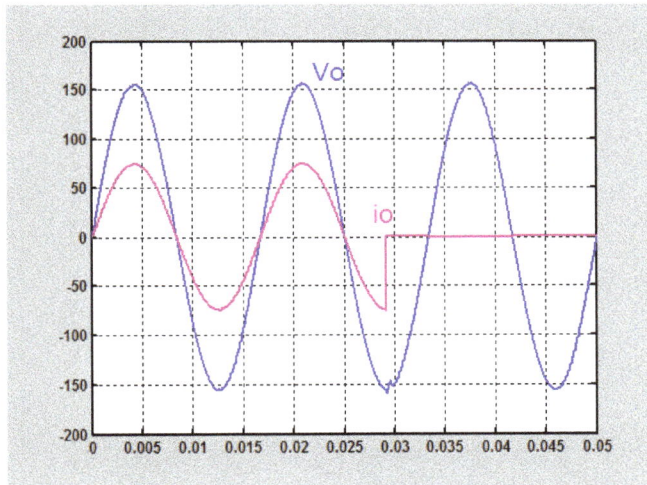

Figure 3. Simulated waveforms under the load suddenly turn off for the classic finite-time convergence control (FTCC) (50 V/div; 10 A/div).

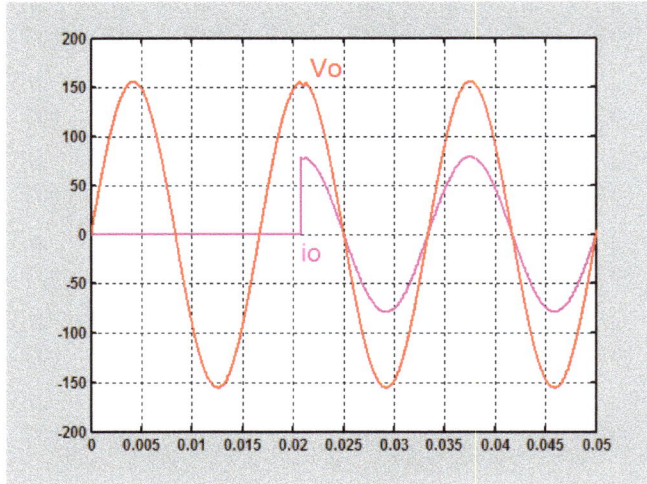

Figure 4. Simulated waveforms in response to the load suddenly turn on for the modified control technology (50 V/div; 10 A/div).

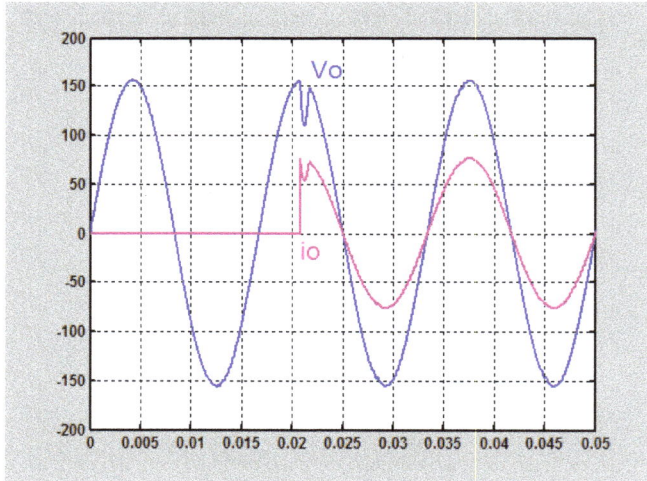

Figure 5. Simulated waveforms in response to the load suddenly turn on for the classic FTCC (50 V/div 10 A/div).

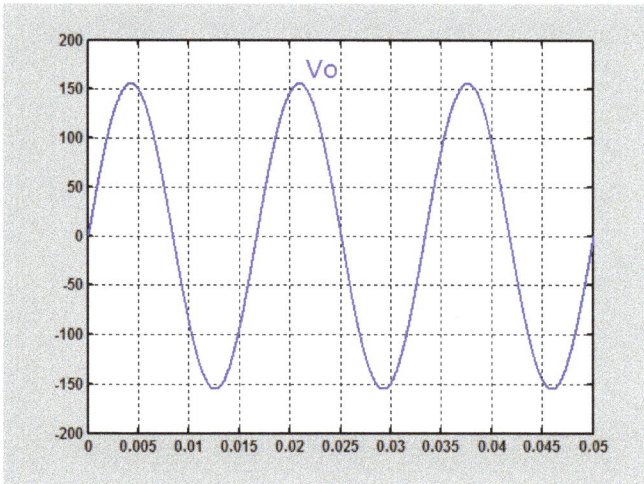

Figure 6. Simulated waveforms in response to the inductor capacitor (LC) variation for the modified control technology (50 V/div).

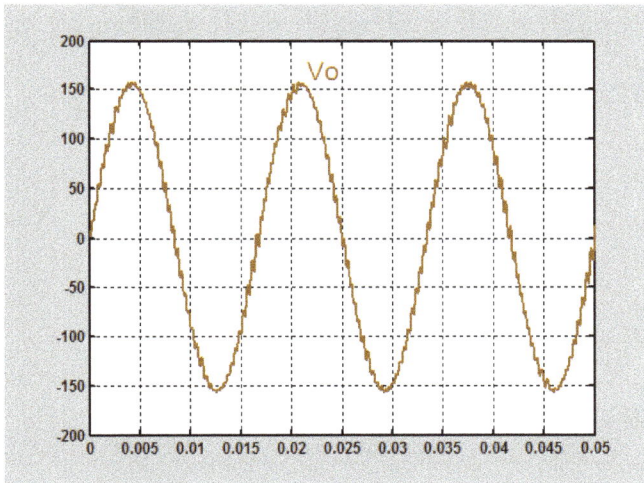

Figure 7. Simulated waveforms in response to the LC variation for the classic FTCC (50 V/div).

Table 2. Simulated output-voltage slump and voltage total harmonic distortion (THD) under a step loading and an inductor capacitor (LC) variation. FTCC—finite-time convergence control.

	Modified Control Technology	
	Step loading (Voltage Slump)	LC variation (Voltage THD)
Simulations	3 V_{rms}	0.08%
	Classic FTCC	
	Step loading (Voltage Slump)	LC variation (Voltage THD)
	32 V_{rms}	10.72%

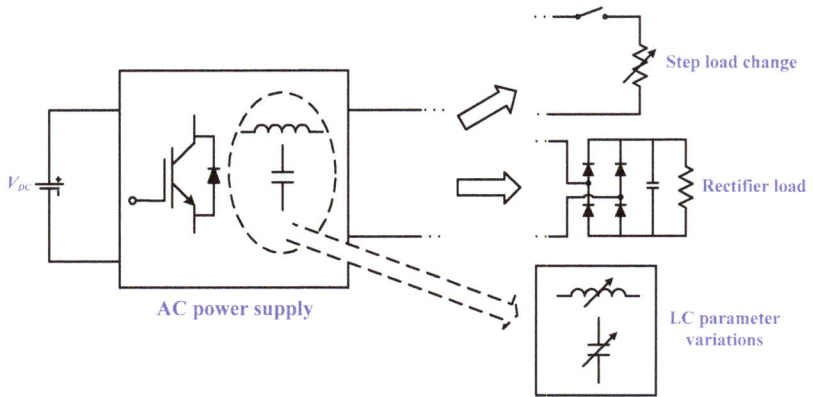

Figure 8. Experimental tests under different load situations.

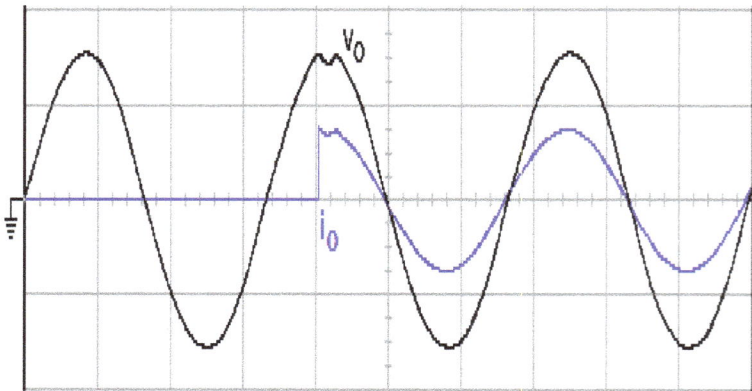

Figure 9. Experimental waveform under a step change in load for the modified control technology (100 V/div; 20 A/div; 5 ms/div).

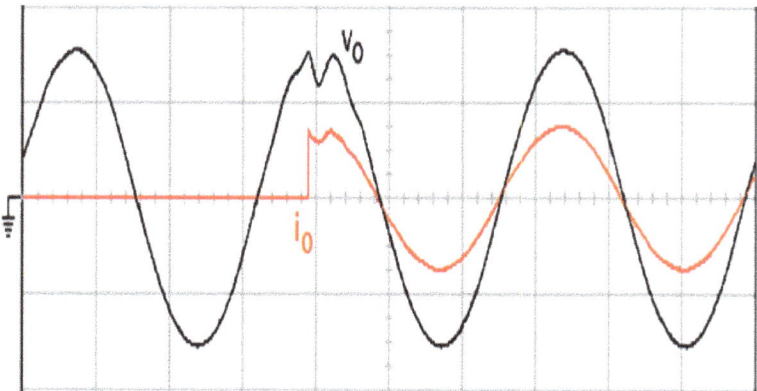

Figure 10. Experimental waveforms under a step change in load for classic FTCC (100 V/div; 20 A/div; 5 ms/div).

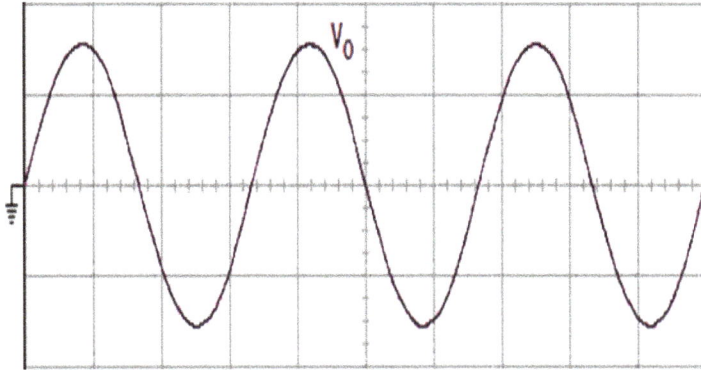

Figure 11. Experimental waveform under a LC variation for the modified control technology (100 V/div; 5 ms/div).

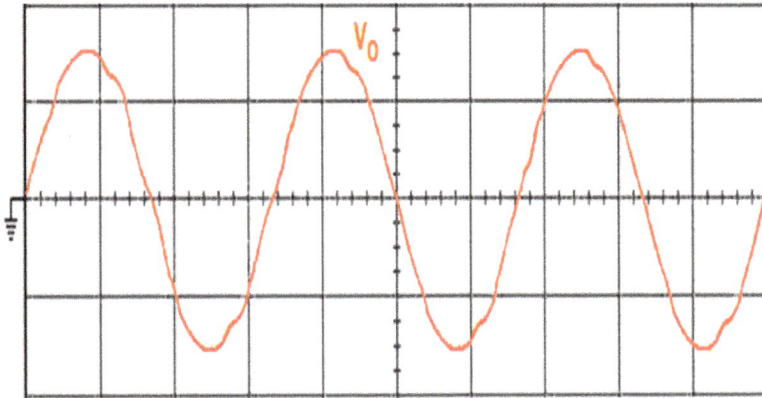

Figure 12. Experimental waveform under an LC variation for the classic FTCC (100 V/div; 5 ms/div).

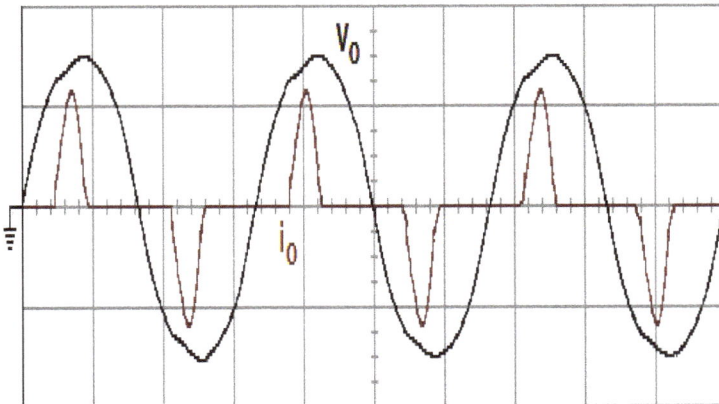

Figure 13. Experimental waveform under a rectifier load for the modified control technology (100 V/div; 25 A/div; 5 ms/div).

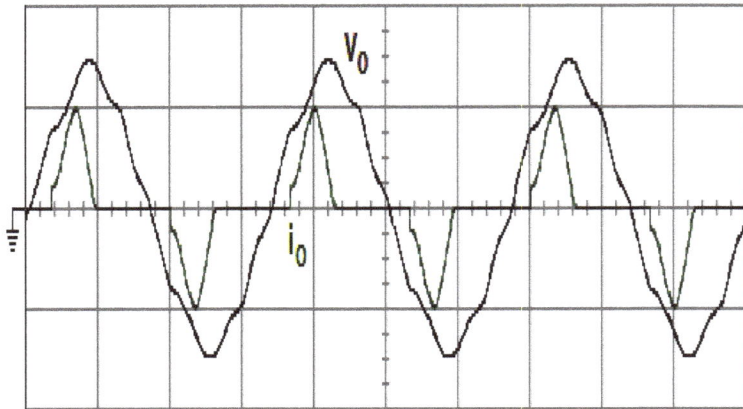

Figure 14. Experimental waveform under a rectifier load for the classic FTCC (100 V/div; 25 A/div; 5 ms/div).

Table 3. Experimental output-voltage slump and voltage THD under a step loading and a rectifier load.

	Modified Control Technology	
	Step loading (Voltage Slump)	Rectifier load (Voltage THD)
Experiments	5 V_{rms}	1.35%
	Classic FTCC	
	Step loading (Voltage Slump)	Rectifier load (Voltage THD)
	30 V_{rms}	8.92%

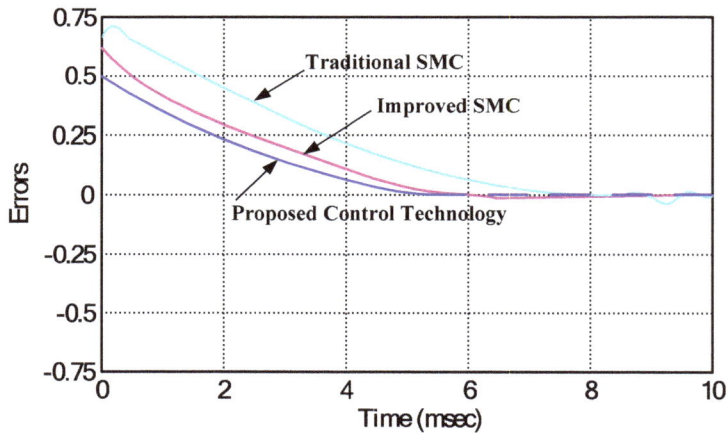

Figure 15. A comparison of tracking errors. SMC—sliding mode control.

5. Discussion

The modified control technology has been proposed for chattering mitigation, steady-state error moderation, and larger perturbation rejection, thus furnishing a good system performance. Nevertheless, for the purpose of the future research, we have reviewed the wide literature in other robust approaches (such as H-infinity controller, mu-synthesis method, and game-theoretic strategy) and high-order SMC (HOSMC) as follows. A discrete-time H-infinity controller is proposed for uninterruptible power supply systems to achieve a nearly zero steady-state error and a mitigate output-voltage distortion caused by non-linear loads. However, this controller is in need of the sophisticated algorithm [23]. Consolidating a mu-synthesis method into the H-infinity controller has attempted to control the islanded micro-grid so that the effects of the parametric uncertainties and external perturbations can be minimized. The digital realization of the mu-synthesis is complex and the resulting waveform yields a conspicuous harmonic distortion, particularly in the strong non-linearity [24]. A game-theoretic strategy is developed for the control of DC (direct current) microgrids, oppressing the harmonic emergence. Although this strategy reveals a good transient and steady state in the face of an abrupt load change and a non-linear load, it is strongly dependent on the exactness of the plant parameters [25]. The traditional (standard) SMC with a linear sliding surface of a single-phase inverter is investigated and scarcely distorts the output-voltage under a linear load. However, the traditional SMC has a long convergence time and chattering effect [26]. The combination of a proportional-resonant and a traditional sliding surface is suggested to improve the transience in the single-phase inverter. The transient behavior can be enhanced, but this methodology results in steady-state errors [27]. The HOSMC approaches are thus used well to control the AC power supply related systems. A grid connected wind system is presented by the multiple-input multiple-output (MIMO) HOSMC, thus regulating the active and reactive powers. This method generates some satisfactory profits, such as the strong robustness to the uncertain perturbations, the mitigation of the chattering, and the finite time convergence of the system states [28]. A backstepping HOSMC strategy is developed for the grid-connected distributed generation (DG) units. It can regulate the inverter output currents and offer the sinusoidal balanced currents to the grid so that the good performance of the distributed generation unit can be obtained in the presence of system uncertainties [29]. A HOSM observer is introduced into the control design of the AC power supply for the sake of rejecting the parametric variations, complex nonlinearities, and external perturbations. The criterion-based Lyapunov function has sternly vouched for the system stability and the virtue of the HOSM observer [30]. As described above, the HOSMC sustains the primordial robustness of the traditional SMC, and simultaneously produces a minor chattering and a preferable convergence precision. Consequently, the HOSMC approaches will stimulate further investigations following this paper in the AC power supply related areas.

6. Conclusions

In this paper, the FTCC with DGPM to design an AC power supply to improve the transient and steady-state behaviors is reported. The FTCC maintains the robustness of the traditional SMC and yields the finite-time convergence of the system state. The employment of the DGPM can estimate the upper bound of the parametric uncertainties and the external perturbations, relieving the overcautious design of the FTCC. The problem of chattering occurring in the FTCC can be solved as a result of the diminution of the overcautious switching gain that enhances the system performance. The modified control technology thus possesses the reachable sliding surface within a finite time, the closed-loop asymptotic stability, and the finite-time convergence to the zero of the tracking errors. On the basis of the theoretic derivation and analysis, simulations, and experimental results, the potency of the modified control technology is successfully attested and very becoming for the use of the single-phase AC power supply. On the other hand, the modified control technology will also offer exciting opportunities to be applied in other circuit structures (such as a high-step-up DC–DC converter, an LED driver, and an online insulation fault detection circuit) [31–33], thus yielding more contributions towards future work.

Author Contributions: E.-C.C. conceived and designed the circuit and developed the methodology. S.-C.Y. and R.-C.W. prepared software resources and set up simulation software. E.-C.C. performed circuit simulations. E.-C.C. carried out the prototype power supply and measured, as well as analyzed, experimental results. E.-C.C. wrote the paper and revised it for submission.

Funding: This research was funded by Ministry of Science and Technology of Taiwan, R.O.C., grant number MOST 107-2221-E-214-006.

Acknowledgments: The authors gratefully acknowledge the financial support of the Ministry of Science and Technology of Taiwan, R.O.C., under project number MOST 107-2221-E-214-006.

Conflicts of Interest: The author declares that there is no conflict of interest regarding the publication of this article.

References

1. Wilamowski, B.M.; Irwin, J.D. *Power Electronics and Motor Drives*; CRC Press: Boca Raton, FL, USA, 2011.
2. Lu, N.J.; Yang, S.F.; Tang, Y. Ripple Current Reduction for Fuel-Cell-Powered Single-Phase Uninterruptible Power Supplies. *IEEE Trans. Ind. Electron.* **2017**, *64*, 6607–6617. [CrossRef]
3. Vaidyanathan, S.; Lien, C.H. *Applications of Sliding Mode Control in Science and Engineering*; Springer: New York, NY, USA, 2017.
4. LWu, G.; Shi, P.; Su, X.J. *Sliding Mode Control of Uncertain Parameter-Switching Hybrid Systems*; Wiley: New York, NY, USA, 2014.
5. Pichan, M.; Rastegar, H. Sliding-Mode Control of Four-Leg Inverter with Fixed Switching Frequency for Uninterruptible Power Supply Applications. *IEEE Trans. Ind. Electron.* **2017**, *64*, 6805–6814. [CrossRef]
6. Gautam, A.R.; Gourav, K.; Guerrero, J.M.; Fulwani, D.M. Ripple Mitigation with Improved Line-Load Transients Response in a Two-Stage DC–DC–AC Converter: Adaptive SMC Approach. *IEEE Trans. Ind. Electron.* **2018**, *65*, 3125–3135. [CrossRef]
7. Altin, N.; Ozdemir, S.; Komurcugil, H.; Sefa, I. Sliding-Mode Control in Natural Frame with Reduced Number of Sensors for Three-Phase Grid-TiedLCL-Interfaced Inverters. *IEEE Trans. Ind. Electron.* **2019**, *66*, 2903–2913. [CrossRef]
8. Lu, K.F.; Xia, Y.Q.; Yu, C.M.; Liu, H.L. Finite-Time Tracking Control of Rigid Spacecraft Under Actuator Saturations and Faults. *IEEE Trans. Autom. Sci. Eng.* **2016**, *13*, 368–381. [CrossRef]
9. Hussian, A.; Zhao, X.D.; Zong, G.D. Finite-Time Exact Tracking Control for a Class of Non-linear Dynamical Systems. *IET Control Theory Appl.* **2017**, *11*, 2020–2027. [CrossRef]
10. Golestani, M.; Mobayen, S.; Tchier, F. Adaptive Finite-Time Tracking Control of Uncertain Non-linear n-Order Systems with Unmatched Uncertainties. *IET Control Theory Appl.* **2016**, *10*, 1675–1683. [CrossRef]
11. Peltoniemi, P.; Nuutinen, P.; Pyrhonen, J. Observer-Based Output Voltage Control for DC Power Distribution Purposes. *IEEE Trans. Power Electron.* **2013**, *28*, 1914–1926. [CrossRef]
12. Wu, X.B.; Liu, Q.; Zhao, M.L.; Chen, M.Y. Monolithic Quasi-Sliding-Mode Controller for SIDO Buck Converter with a Self-Adaptive Free-wheeling Current Level. *J. Semicond.* **2013**, *34*, 1–7. [CrossRef]
13. Shen, L.; Lu, D.D.; Li, C. Adaptive Sliding Mode Control Method for DC-DC Converters. *IET Power Electron.* **2015**, *8*, 1723–1732. [CrossRef]
14. Liu, S.F.; Lin, Y. *Advances in Grey Systems Research*; Springer: Heidelberg/Berlin, Germany, 2010.
15. Deng, J.L. Introduction to grey system theory. *J. Grey Syst.* **1989**, *1*, 1–24.
16. Wang, M.H.; Tsai, H.H. Fuel cell fault forecasting system using grey and extension theories. *IET Renew. Power Gener.* **2012**, *6*, 373–380. [CrossRef]
17. Wang, M.D.; Wang, X.K.; Su, X.J.; Li, X.Y.; Yang, Y.H. Generator governing system based on grey prediction and extension control. *IET Gener. Transm. Distrib.* **2017**, *11*, 3776–3782. [CrossRef]
18. Truong, D.Q.; Ahn, K.K.; Trung, N.T. Design of An Advanced Time Delay Measurement and A Smart Adaptive Unequal Interval Grey Predictor for Real-Time Nonlinear Control Systems. *IEEE Trans. Ind. Electron.* **2013**, *60*, 4574–4589. [CrossRef]
19. Samet, H.; Mojallal, A. Enhancement of Electric ARC Furnace Reactive Power Compensation Using Grey-Markov Prediction Method. *IET Gener. Transm. Distrib.* **2014**, *8*, 1626–1636. [CrossRef]
20. Ahmad, A.A.; Abrishamifar, A.; Farzi, M. A New Design Procedure for Output LC Filter of Single Phase Inverters. In Proceedings of the International Conference Power Electronics and Intelligent Transportation System, Shenzhen, China, 13–14 November 2010; pp. 86–91.
21. Dahono, P.A.; Purwadi, A.; Qamaruzzaman. An LC filter Design Method for Single-Phase PWM Inverters. In Proceedings of the International Conference Power Electronics and Drive Systems, Singapore, 21–24 February 1995; pp. 571–576.

22. Kim, H.S.; Sul, S.K. A Novel Filter Design for Output LC Filters of PWM Inverters. *J. Power Electron.* **2011**, *11*, 74–81. [CrossRef]

23. Darvishzadeh, S.; Rahmati, A.; Abrishamifar, A. Comparative Study of Different Switching Surfaces for Sliding Mode Control of a 40 kVA Single-phase UPS Inverter. *Int. J. Comput. Electr. Eng.* **2012**, *4*, 933–936. [CrossRef]

24. Aamir, M.; Kalwar, K.A.; Mekhilef, S. Proportional-Resonant and Slide Mode Control for Single-Phase UPS Inverter. *Electr. Power Compon. Syst.* **2017**, *45*, 11–21. [CrossRef]

25. Ribas, S.P.; Maccari, L.A.; Pinheiro, H.; Oliveira, R.C.; Montagner, V.F. Design and Implementation of a Discrete-time H-infinity Controller for Uninterruptible Power Supply Systems. *IET Power Electron.* **2014**, *7*, 2233–2241. [CrossRef]

26. Bevrani, H.; Feizi, M.R.; Ataee, S. Robust Frequency Control in an Islanded Microgrid: H-infinity and Mu-Synthesis Approaches. *IEEE Trans. Smart Grid* **2016**, *7*, 706–717. [CrossRef]

27. Ekneligoda, N.C.; Weaver, W.W. Game-Theoretic Cold-Start Transient Optimization in DC Microgrids. *IEEE Trans. Ind. Electron.* **2014**, *61*, 6681–6690. [CrossRef]

28. Valenciaga, F.; Fernandez, R.D. Multiple-Input–Multiple-Output High-Order Sliding Mode Control for a Permanent Magnet Synchronous Generator Wind-Based System with Grid Support Capabilities. *IET Renew. Power Gener.* **2015**, *9*, 925–934. [CrossRef]

29. Dehkordi, N.M.; Sadati, N.; Hamzeh, M. A Robust Backstepping High-Order Sliding Mode Control Strategy for Grid-Connected DG Units with Harmonic/Interharmonic Current Compensation Capability. *IEEE Trans. Sustain. Energy* **2017**, *8*, 561–572. [CrossRef]

30. Chen, D.; Jun, Y.; Wang, Z.; Li, S.H. Universal Active Disturbance Rejection Control for Non-linear Systems with Multiple Disturbances via a High-Order Sliding Mode Observer. *IET Control Theory Appl.* **2017**, *11*, 1194–1204.

31. Cheng, C.A.; Cheng, H.L.; Chang, C.H.; Chang, E.C.; Yang, F.L. Design and Implementation of a Novel High-Step-Up DC-DC Converter. *Appl. Mech. Mater.* **2013**, *284–287*, 2498–2501.

32. Cheng, C.A.; Chang, E.C.; Tseng, C.S.; Chung, T.Y. A Novel High-Power-Factor LED-Lamp Driver Based on a Single-Stage Power Conversion. In Proceedings of the 2014 International Symposium on Computer, Consumer and Control, Taichung, Taiwan, 10–12 June 2014; pp. 1287–1290.

33. Liu, Y.C.; Chang, E.C.; Lin, Y.L.; Lin, C.Y. A Novel Online Insulation Fault Detection Circuit for DC Power Supply Systems. *Int. J. Smart Grid Clean Energy* **2018**, *7*, 64–73. [CrossRef]

electronics

MDPI

Article

A Mixed Deep Recurrent Neural Network for MEMS Gyroscope Noise Suppressing

Changhui Jiang [1,2], Yuwei Chen [2], Shuai Chen [1,*], Yuming Bo [1], Wei Li [3], Wenxin Tian [3] and Jun Guo [1]

[1]   School of automation, Nanjing University of Science and Technology, Nanjing 210094, China; Chiang_changhui@outlook.com (C.J.); byming@mail.njust.edu.cn (Y.B.); guojun1136@163.com (J.G.)
[2]   Centre of Excellence in Laser Scanning Research, Finnish Geospatial Research Institute (FGI), Geodeetinrinne 2, FI-02431 Kirkkonummi, Finland; yuwei.chen@nls.fi
[3]   Key Laboratory of Quantitative Remote Sensing Information Technology, Chinese Academy of Sciences (CAS), Beijing 100094, China; liwei@aoe.ac.cn (W.L.); tianwenxin@aoe.ac.cn (W.T.)
*   Correspondence: c1492@163.com; Tel.: +86-138-139-15826

Received: 8 December 2018; Accepted: 14 January 2019; Published: 4 February 2019

Abstract: Currently, positioning, navigation, and timing information is becoming more and more vital for both civil and military applications. Integration of the global navigation satellite system and /inertial navigation system is the most popular solution for various carriers or vehicle positioning. As is well-known, the global navigation satellite system positioning accuracy will degrade in signal challenging environments. Under this condition, the integration system will fade to a standalone inertial navigation system outputting navigation solutions. However, without outer aiding, positioning errors of the inertial navigation system diverge quickly due to the noise contained in the raw data of the inertial measurement unit. In particular, the micromechanics system inertial measurement unit experiences more complex errors due to the manufacturing technology. To improve the navigation accuracy of inertial navigation systems, one effective approach is to model the raw signal noise and suppress it. Commonly, an inertial measurement unit is composed of three gyroscopes and three accelerometers, among them, the gyroscopes play an important role in the accuracy of the inertial navigation system's navigation solutions. Motivated by this problem, in this paper, an advanced deep recurrent neural network was employed and evaluated in noise modeling of a micromechanics system gyroscope. Specifically, a deep long short term memory recurrent neural network and a deep gated recurrent unit–recurrent neural network were combined together to construct a two-layer recurrent neural network for noise modeling. In this method, the gyroscope data were treated as a time series, and a real dataset from a micromechanics system inertial measurement unit was employed in the experiments. The results showed that, compared to the two-layer long short term memory, the three-axis attitude errors of the mixed long short term memory–gated recurrent unit decreased by 7.8%, 20.0%, and 5.1%. When compared with the two-layer gated recurrent unit, the proposed method showed 15.9%, 14.3%, and 10.5% improvement. These results supported a positive conclusion on the performance of designed method, specifically, the mixed deep recurrent neural networks outperformed than the two-layer gated recurrent unit and the two-layer long short term memory recurrent neural networks.

Keywords: global navigation satellite system (GNSS); inertial navigation system (INS); long short term memory (LSTM); gated recurrent unit (GRU); microelectronics system (MEMS)

---

1. Introduction

With the booming of location based services (LBS), positioning, navigation, and timing (PNT) information is more essential than at any time in human history, since more and more smart devices relies on PNT information [1]. Currently, the global navigation satellite system (GNSS) has been the most widely used PNT information provider and generator, due to its easy access, low cost, and high accuracy. Broadly speaking, the GNSS refers to all satellite based navigation systems, including global and regional systems. Among them, the USA Global Positioning System (GPS), China BeiDou Navigation System (BDS), Europe Galileo Satellite Navigation System (Galileo), and the Russia Global Navigation Satellite System (GLONASS) are capable of global coverage, and other regional systems, for instance Japan's Quasi-Zenith Satellite System (QZSS) and the Indian Regional Navigation Satellite System (IRNSS), offer an augmentation of GPS for performance enhancement in specific regions [1–3]. Generally, their working principles are similar, and the details are as follows: (1) firstly, the satellites in orbit broadcast navigation signals to the Earth, and the signals are modulated with information of the satellites' orbit description parameters and some other information; (2) secondly, the user receives the broadcast signals and de-modulates the information, which can be employed to obtain the distance between the user and satellites; (3) thirdly, with at least four satellites in view, the PNT information can be determined precisely using a least-square algorithm or Kalman filter [1–5]. The advantages of GNSS are summarized as: (1) GNSS is able to provide precise navigation solutions at low cost, since a handheld chip receiver is cheap and sufficient for common applications; (2) GNSS is an all-weather navigation system covering the earth, and its positioning accuracy does not diverge over time. However, apart from these advantages, it also has some drawbacks limiting its further application: (1) firstly, the satellites are far away from the Earth, thus, the signals are pretty weak when they reach the Earth; (2) secondly, GNSS civil signal structure is open to the public, which makes GNSS extremely sensitive to interference and spoofing; (3) thirdly, temporary signal blockages or obstruction can also render the GNSS receiver unavailable to the satellite signals [6–10]. A standalone GNSS is not sufficient to provide seamless PNT information, thus they are commonly integrated with an inertial navigation system (INS) to provide ubiquitous navigation solutions [6–10]. While the GNSS is unavailable, the INS outputs the positioning information for users during the signal outage.

INS is another navigation system capable of providing position, velocity, and attitude information. An INS is constructed through processing raw data or signals from the inertial measurement unit (IMU). Commonly, an IMU consists of three accelerometers and gyroscopes. Positioning errors divergence is usually caused by the noise contained in raw signals from the gyroscopes and accelerometers. Recently, due to the low cost and small size of the advanced micro-mechanics system (MEMS) manufacturing technology, the MEMS IMU has become more popular in the community for developing low cost and highly accurate GNSS/INS integrated navigation systems. However, as the MEMS IMU experiences more complex errors and noises, it is of great significance to develop a noise modeling method for the MEMS IMU [11–19], especially for improving positioning accuracy during GNSS signal outages.

In INS, gyroscopes play an important role in INS positioning accuracy, thus, past works have mostly focused on modeling and suppressing the noise of the MEMS gyroscopes [11–19]. Various methods have been proposed and evaluated in MEMS gyroscope noise analysis and modeling; and basically, the methods can be classified into two approaches: statistical method and artificial intelligence method. In the statistical methods, Allan Variance (AV) and Auto Regressive Moving Average (ARMA) are the most popular. AV was first employed in MEMS IMU noise analysis and errors description in 2004 [19]. In the AV method, five basic parameters are introduced to describe the gyroscopes' and accelerometers' noise, and the parameters are termed as: quantization noise, angle random walk, bias instability, rate random walk, and rate ramp [19–24]. ARMA is another method for MEMS gyroscope noise modeling and compensation, in which the raw data are treated as time series. Variants of ARMA have also been proposed to furtherly improve the performance [25–29]. Moreover, artificial intelligence methods, such as support vector machines (SVM) and neural networks (NN), have also been employed in this application to obtain better de-noising performance [30–32]. The results demonstrate the effectiveness of these methods in this application. However, both of the two solutions have some drawbacks, the statistical method usually has fixed parameters, which are not sufficient for certain applications; the artificial methods usually have limited ability to learn the model, due to their simple structures.

Recently, Deep Learning (DL) has gained a boom and performed excellently in various applications including image processing, Nature Language Processing (NLP) and sequential signal processing [30–37]. In aspects of time series processing, a recurrent neural network (RNN) was always the most feasible selection [30–37]. A common RNN was not sufficient, thus, variants of RNN were proposed for enhancing the performance. Among the variants, Long Short Term Memory (LSTM) and the Gated Recurrent Unit (GRU) were most popular. LSTM-RNN and GRU-RNN both obtained excellent performance in NLP [30–37]. In addition, in our previous paper, LSTM was employed and compared in MEMS gyroscope de-noising [8]. With fixed or identical length of training examples, LSTM had better training accuracy, but GRU had better convergence efficiency for its unique design [8]. Commonly, GRU was designed with less parameters than LSTM, and this made GRU coverage faster and quickly than LSTM in training procedures.

Inspired by the multi-layer RNN design scheme, a new architecture combing LSTM and GRU together was explored for MEMS gyroscope noise modeling in the paper. As aforementioned, since the GRU and LSTM had different characteristics, it was meaningful to explore the mixed LSTM and GRU in this application. Specifically, in this paper, two multi-layer RNNs with different architectures (LSTM–GRU: first layer, LSTM; second layer, GRU. GRU–LSTM: first layer, GRU; second layer: LSTM) of LSTM and GRU combination were investigated.

In this method, a GRU unit was substituted by a LSTM unit in a two layer GRU-RNN, thus, the method was expected to combine the advantages from the LSTM and GRU. An MEMS IMU dataset was collected to evaluate the proposed method, and compare the results with a common multi–layer GRU-RNN and multi-layer LSTM-RNN. Firstly, LSTM–GRU and GRU–LSTM were compared to select the proper structure for this application. Secondly, the new method was compared with a multi-layer LSTM-RNN and multi-layer GRU-RNN for a more specific analysis of performance. Finally, the standard deviation of the filtered signals and the attitude errors were presented. We thought the contributions of this paper could be summarized as:

(1) It was the first time a mixed LSTM and GRU method has been applied to MEMS gyroscope noise modeling, which might be an inspiration for applying DL in MEMS IMU de-noising.

(2) It was a bright idea to develop a mixed multi-layer RNN; detailed analysis of the multi-layer LSTM, multi-layer GRU, LSTM–GRU, and GRU–LSTM were presented and compared, which could provide valid reference while selecting proper methods for MEMS gyroscope noise modeling.

The remainder of this paper is organized as follows: Section 2 describes the structures and the equations of the employed RNN, including the LSTM unit, GRU unit, and the mixed LSTM and GRU. Section 3 introduces the experiments, results, and analysis of these methods. The remaining sections are the discussion, conclusion, acknowledgements, and the references.

2. Methods

In this section, the basic structure and detailed mathematical equations are listed. This section is divided into four subsections. Section 2.1 is the basic introduction of the LSTM unit, Section 2.2 is about the GRU, Section 2.3 presents the combination of LSTM and GRU.

2.1. Long Short Term Memory (LSTM)

As is well-known, LSTM is built using a unique 'gate' structure. Figure 1 shows the basic components of a LSTM. 'Forget,' 'Input,' and 'Output' gates work cooperatively to accomplish the function of a LSTM unit and control the information flow. As presented in Figure 1, from left to the right, the first component is the 'forget' gate, and a sigmoid function $\sigma(\cdot)$ is employed in this gate to decide what information will be memorized from the previous state cell. The details of this procedure are listed as the following Equation (1).

$$f_t = \sigma\left(W_f \cdot [h_{t-1}, x_t] + b_f\right) \tag{1}$$

where, $\sigma(\cdot)$ is the sigmoid function, W_f and b_f are the parameters that will be determined after training, h_{t-1} is the hidden state at time epoch $t-1$, and x_t is the input vector at time epoch t. Vector f_t is the output of the sigmoid function.

The inputs of the function are the hidden state from the previous LSTM unit and input vector. Outputs of the functions are values ranging from 0 to 1, which correspond to each number in the cell state from the previous LSTM unit. The values represent the forgetting degree of each number in the previous cell state C_{t-1}. A value of '1' means 'completely keeping this,' and, oppositely, a value of '0' means 'completely forgetting or excluding.

After the "forget" gate, the following is the 'input' gate, which controls the input and decides what part of the new information will be stored in the current cell state. The procedure is operated using the following two functions, Equations (2) and (3). Equation (2) is a sigmoid function similar to Equation (1). This function is employed to decide the updating degree of each number in the input vector. Equation (3) is a *tanh* layer, which outputs a new cell state $\widetilde{C}_t$. Later, the new, hidden $\widetilde{C}_t$ is multiplied with the vector i, and then added to the current cell state.

$$i_t = \sigma(W_i \cdot [h_{t-1}, x_t] + b_i) \tag{2}$$

$$\widetilde{C}_t = \tanh(W_C \cdot [h_{t-1}, x_t] + b_C) \tag{3}$$

where $\sigma(\cdot)$ is a sigmoid function, W_i, b_i, W_C, and b_C are the parameters will be determined through training procedure, h_{t-1} is the hidden state at time $t-1$, and x_t is the input vector.

Thirdly, an 'output' gate is employed to decide and control the outputs. This 'gate' is also composed of two functions: a sigmoid function and a *tanh* function. The details are listed as Equations (4) and (5). The output of the sigmoid function is o_t, which decides the outputs of the hidden state. The cell state is then put through a *tanh* function and multiplied with the vector o_t, deciding the outputs of the LSTM unit.

$$o_t = \sigma(W_o \cdot [h_{t-1}, x_t] + b_o) \tag{4}$$

$$h_t = o_t * tanh(C_t) \tag{5}$$

where, W_o and b_o are the parameters determined during the training, and C_t is the cell state at time t.

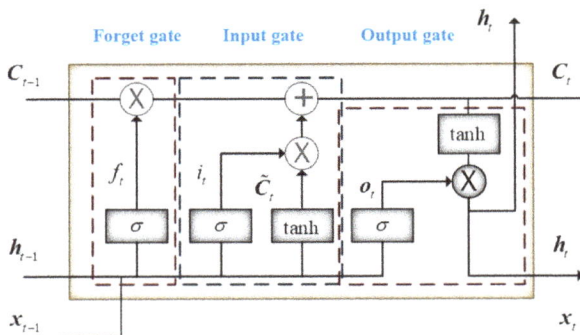

Figure 1. Basic structure of a long short term memory (LSTM) unit.

2.2. Gated Recurrent Unit (GRU)

The gated recurrent unit is another popular variant of the common RNN, and was first proposed by Cho in 2004 [32]. In a GRU, the information flow is also controlled and monitored based on a 'gate' structure, however, a GRU has no separate state cells. A GRU basic structure is shown in Figure 2. In the figure, h_{t-1} is the hidden state at time $t-1$, and h_t is the hidden state at time epoch t. The relationship between h_{t-1} and h_t is as Equation (6):

$$h_t = (1 - z_t) * h_{t-1} + z_t * \tilde{h}_t \tag{6}$$

where $\tilde{h}_t$ is the candidate activation or hidden state, and the updating gate z_t decides how much the unit updates its hidden state, which is as Equation (7):

$$z_t = \sigma(W_z \cdot [h_{t-1}, x_t]) \tag{7}$$

where $\sigma(\cdot)$ is a sigmoid function, and determining the degree of the new hidden state will be added to the hidden state at time epoch t. In above, W_z is the parameters which will be determined after training.

In addition, the new or candidate hidden state $\tilde{h}_t$ calculation is as Equation (8):

$$\tilde{h}_t = tanh(W \cdot [r_t * h_{t-1}, x_t]) \tag{8}$$

where r_t is a set of reset gates, and when r_t is close to 0, the unit acts as forgetting the previously computed state. r_t is calculated as:

$$r_t = \sigma(W_r \cdot [h_{t-1}, x_t]) \tag{9}$$

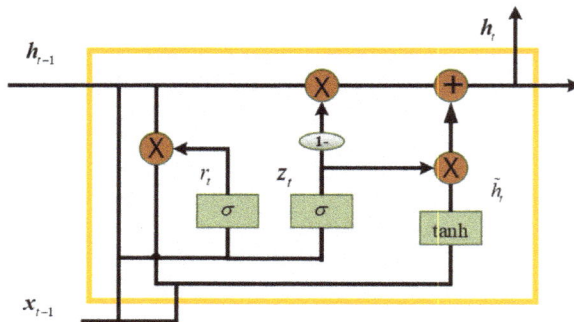

Figure 2. Basis structure of a gate recurrent unit (GRU).

2.3. Mixed LSTM and GRU

As presented in Figures 1 and 2, the LSTM and GRU are just a single unit. A deep LSTM-RNN or deep GRU-RNN is set up as Figure 3, the LSTM and GRU units are assembled and connected in the time domain, and the parameters propagation is also illustrated in detail. The output is determined by a LSTM or GRU sequence, thus, long term memory could also affect the current epoch output. Furtherly, Figure 4 presents the mixed LSTM and GRU deep RNN structures, Figure 4a shows the LSTM–GRU. In this structure, the cell state and hidden state are converted to the LSTM unit at the next epoch. The hidden state is also converted to the parallel GRU unit, and the hidden state propagates among the GRU units. In the GRU–LSTM structure (Figure 4b), the hidden state of GRU is converted to the next GRU unit and the parallel LSTM unit. Since the GRU has no cell state, in the mixed LSTM–GRU structure the cell state only propagates among the LSTM units.

(a)

(b)

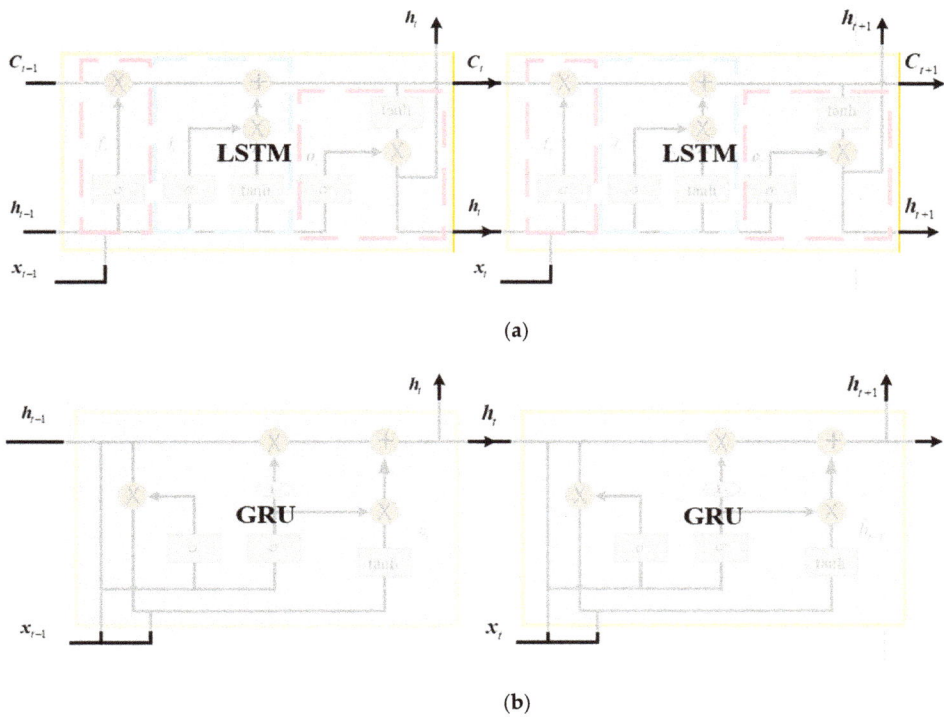

Figure 3. Deep LSTM and GRU with single layers. (**a**) A single layer LSTM; (**b**) a single layer GRU.

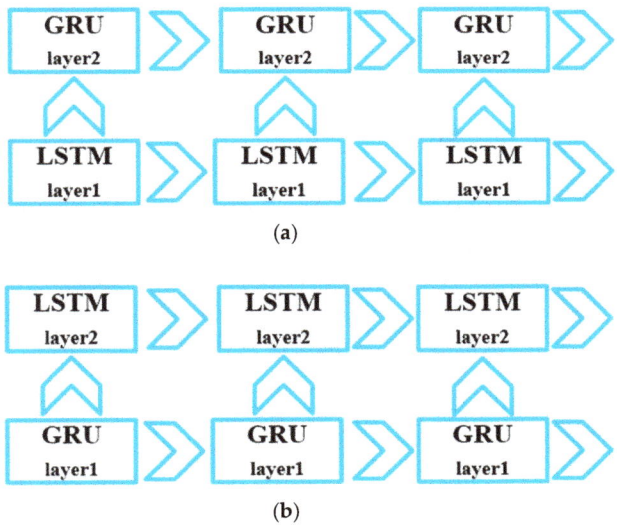

(a)

(b)

Figure 4. Structures of the mixed LSTM and GRU. (**a**) LSTM–GRU; (**b**) GRU–LSTM.

3. Results

This section introduces the experimental setup and the results. Figure 5 presents the data collecting procedure; a MEMS IMU (MT Microsystem Company, Hubei, China) is employed, and the details are listed in Table 1 [33]. The MEMS IMU is same model that is employed in our previous paper, but they are not from the same batch [8]. Thus, they have similar parameters, but are actually different after precise calibration. This difference is caused by the MEMS manufacturing technology.

Figure 5 presents the data collecting equipment. The power supply delivers 12 V and 0.11 A while the MEMS IMU is connected. A computer is also connected to the MEMS IMU through a USB 2.0 cable. Control software is run by the computer to monitor the data collecting procedure, obtain, and store the data. The sampling frequency is set at 200 Hz, and the collecting time length is approximately 600 s.

The remainder of this section is divided into three sections: Section 3.1 illustrates the formulae of the gyroscope output data errors, the structure of the input, and output data for training and testing. Section 3.2 details the investigation of the training data length on the proposed mixed LSTM and GRU method, since the GRU had better performance with less training data length compared to the LSTM. The aim of the Section 3.2 was to explore the performance of the mixed LSTM and GRU RNN, as compared with multi-layer LSTM and multi-layer GRU. In Section 3.3, the performance of the new method was compared with the two-layer LSTM and two-layer GRU to provide a detailed description of the proposed method. Attitude errors are also presented for further comparison of these methods.

Table 1. Specifications of MSI3200 IMU.

MEMS IMU	Gyroscope	range	$\pm300\,°/s$
		Bias stability (1 σ)	$\leq10\,°/h$
		Bias stability (Allan)	$\leq2\,°/h$
		Angle random walk	$\leq10\,°/\sqrt{h}$
	Accelerometer	range	$\pm15\,g$
		Bias stability (1 σ)	0.5 mg
		Bias stability (Allan)	0.5 mg
	Power consumption		1.5 W
	Weight		250 g
	Size		70 mm $\times$ 54 mm $\times$ 39 mm
	Sampling rate		400 Hz

Figure 5. MEMS Gyroscope data collecting procedure.

3.1. Input Data and Training

As illustrated in Figure 5, the gyroscope dataset was collected. The bias was calculated using the mean values of the collected data. After subtracting the bias, the processed dataset was labeled as $X = [x_1, x_2, x_3, \ldots, x_N]$. The subscript N was termed as the number of input gyroscope samples. In the experiments, the dataset was divided into two parts: training part ($X_{training}$) and testing part ($X_{testing}$). The training dataset was employed to train the model, and the testing dataset was utilized to evaluate the performance of the trained model. The input vector of the RNNs could be described as:

$$input_i = [x_i, x_{i+1}, \ldots, x_{i+step}], i \in [1, N_{training} - step] \tag{10}$$

where $input_i$ is the input vector of the RNNs and the variable $step$ is the length of the $input_i$ vector. The output vector $output_i$ is described by:

$$output_i = [x_{i+step+1}], i \in [1, N_{training} - step] \tag{11}$$

Equations (10) and (11) give the dataset for the training procedure, and the dataset in the testing step was similar to that of training procedure.

RNNs were trained using the errors back propagation method (BP). Since the RNNs were designed to process time series datasets, the BP method is termed BPTT (back-propagation through time) [36–38].

3.2. Comparison of LSTM–GRU and GRU–LSTM

As aforementioned in Section 2.3, there were two different architectures in mixed LSTM and GRU. This section aimed to compare these two architectures in aspects of training loss and prediction accuracy. The date lengths of the training dataset and testing dataset were 1000 and 100,000, respectively. The learning rate was 0.01 for both, the hidden unit was 1, and the training epoch was 50. Figures 6–8 present the training loss comparison of the LSTM–GRU and GRU–LSTM in the three-axis MEMS gyroscope de-noising. In addition, training loss means the errors between the predicted values and the real signal values were not included in the training dataset.

In these figures (Figures 6–8), the red line represents the LSTM–GRU training loss, and the blue line represents the GRU–LSTM results. Figure 6 shows the x axis MEMS gyroscope results; the GRU–LSTM and LSTM–GRU both converged within 50 training epochs, but the GRU–LSTM delivered a lower convergence speed with smaller training loss. In Figure 7, the GRU–LSTM and LSTM–GRU seemed not to converge, while the LSTM–GRU had a better performance in reducing training loss. For the z axis MEMS gyroscope, LSTM–GRU converged fast to a stable value, while the GRU–LSTM did not converge within the set training epoch. We thought the difference was caused by the different architectures between LSTM–GRU and GRU–LSTM. LSTM-RNN had more parameters, which needed to be determined during the training procedure, When the LSTM was placed on the second layer, it was not sufficient for LSTM unit training. Overall, the LSTM–GRU was more feasible for this application, compared with the GRU–LSTM. Specifically, the prediction results are not presented, since the GRU–LSTM was not well trained with the settings.

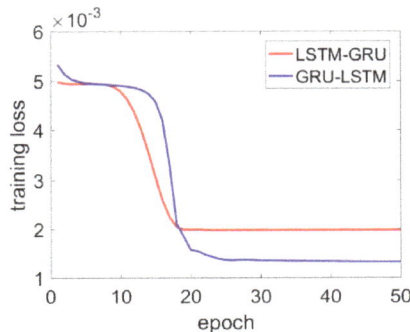

Figure 6. LSTM–GRU and GRU–LSTM training loss for the x axis gyroscope.

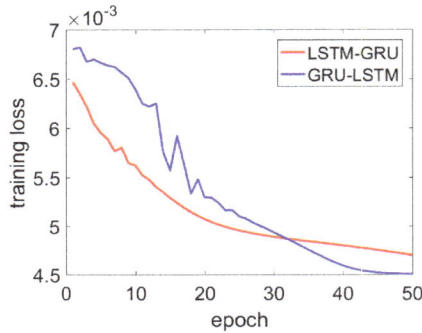

Figure 7. LSTM–GRU and GRU–LSTM training loss for the y axis gyroscope.

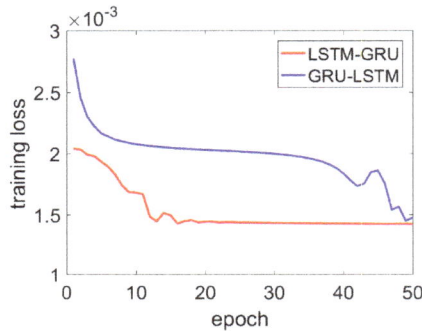

Figure 8. LSTM–GRU and GRU–LSTM training loss for the z axis gyroscope.

3.3. Comparison of LSTM–GRU, Two-Layer LSTM, and Two-Layer GRU

This sub-section presents the comparison results from the two-layer LSTM, two-layer GRU, and the mixed LSTM–GRU. Tables 2–4 show the training loss, standard deviation of the prediction results, and standard deviation of the raw MEMS gyroscope signals. In particular, the structure of the LSTM–GRU is shown in Figure 4. For the x axis gyroscope, the LSTM–GRU delivered the smallest training loss, however, the standard deviations of the de-noised signals were minor. In aspects of the y axis results, the training loss decreased by 12.2% and 14.7%, compared with that of the two-layer LSTM and two-layer GRU. However, the standard deviation of the de-noised signals did not show a significant improvement. For the z axis gyroscope results, the differences in the training loss between the two-layer LSTM, two-layer GRU, and LSTM–GRU were trivial.

In addition, Figures 9–12 present the detailed training losses during the training procedure. In these figures, the red line represents the training loss of the LSTM, the blue line shows the GRU results, and the last green line shows the LSTM–GRU training loss. Specifically, in Figure 9, the two-layer GRU and two-layer LSTM had better performance than the LSTM–GRU. For the LSTM–GRU, the training loss remained almost unchanging, and it converged quickly from the 10th to the 20th training epoch. In Figure 10, the LSTM–GRU outperformed the LSTM and GRU. Figures 11 and 12 show the z axis gyroscope de-noised results. Figure 12 shows a magnified picture of the results from the 5th to the 20th epoch.

Basically, LSTM–GRU showed a slower convergence speed, especially the training epochs from 1 to 10. The phenomenon is obvious in Figures 9 and 11. However, in Figure 10, the training loss of the LSTM–GRU was always below the two-layer LSTM and two-layer GRU. Moreover, the LSTM–GRU delivered smaller training loss than the two-layer LSTM and two layer GRU for the de-noised y axis and z axis gyroscope signals.

Table 2. Standard deviation of gyroscope outputs (two-layer LSTM-RNN).

	X (degree/s)	Y (degree/s)	Z (degree/s)
Training loss	0.00132	0.00534	0.00139
LSTM-RNN	0.060	0.037	0.025
Original signals	0.069	0.083	0.047

Table 3. Standard deviation of gyroscope outputs (two-layer GRU-RNN).

	X (degree/s)	Y (degree/s)	Z (degree/s)
Training loss	0.00136	0.0055	0.00142
LSTM-RNN	0.059	0.034	0.026
Original signals	0.069	0.083	0.047

Table 4. Standard deviation of gyroscope outputs (mixed LSTM–GRU RNN).

	X (degree/s)	Y (degree/s)	Z (degree/s)
Training loss	0.00127	0.00469	0.00134
LSTM-RNN	0.060	0.035	0.0246
Original signals	0.069	0.083	0.047

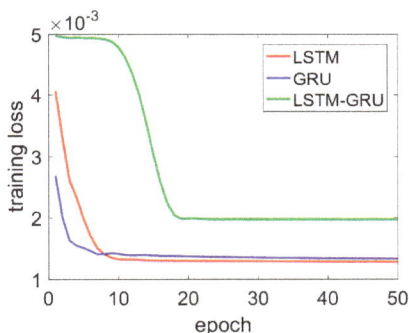

Figure 9. LSTM, GRU, and LSTM–GRU training loss comparison for x axis gyroscope.

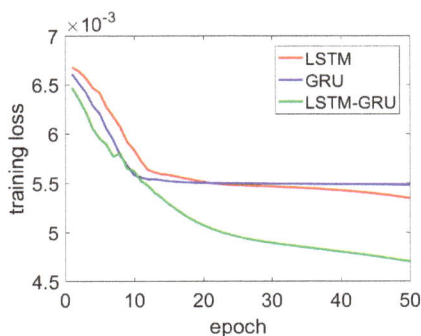

Figure 10. LSTM, GRU, and LSTM–GRU training loss comparison for x axis gyroscope.

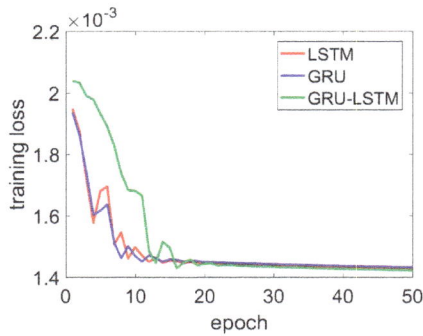

Figure 11. LSTM, GRU, and LSTM–GRU training loss comparison for z axis gyroscope.

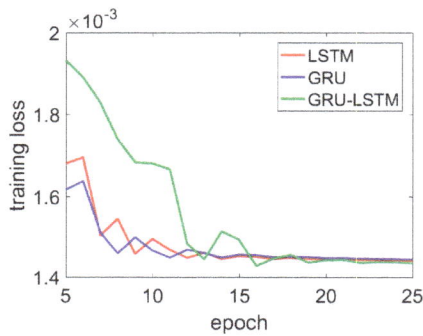

Figure 12. Zooming out of Figure 11 from the 5th to 25th epoch.

Furthermore, Table 5 presents the attitude errors comparison of the three recurrent neural networks (60 s). In Table 5, the three axes referred to pitch, roll, and yaw angles respectively. From the results, three major conclusions were obtained:

(1) There was an obvious improvement in the attitude errors for all the three deep neural networks. The two-layer LSTM performed 64.4%, 49.3%, and 53.3% improvements in attitude errors, the two-layer GRU performed 56.3%, 54.5%, and 47.9% decreases in attitude errors, and the attitude errors of LSTM–GRU decreased by 72.2%, 69.3%, and 58.4%.

(2) Specifically, for the x axis gyroscope data, LSTM–GRU had a large training loss, but the LSTM–GRU still showed 7.8% and 15.9% improvements compared with the two-layer LSTM and two-layer GRU. The minor difference of the standard deviation of the de-noised signals may account for this.

Table 5. Attitude errors comparison.

	X (degree)	Y (degree)	Z (degree)
two-layer LSTM	0.136	0.240	0.184
two-layer GRU	0.167	0.215	0.205
LSTM–GRU	0.104	0.145	0.164
Original signals	0.382	0.473	0.394

4. Discussion

In this paper, the influence of data length on the training performance was not presented and analyzed, as we were limited by the computer computation capacity. Longer training datasets might improve the performance of the deep recurrent neural network.

In the experiment, only static data were employed to evaluate the proposed method; a trajectory from field testing might be more feasible for sufficient testing.

As we were limited by computer capacity, only two-layer LSTM or GRU were employed and implemented in this paper. It may be meaningful to explore the LSTM or GRU with more layers.

5. Conclusions

In this paper, a proposed artificial intelligence method was employed and evaluated in MEMS three-axis gyroscope signal de-noising. Through the experiments, the following conclusions were obtained:

(1) Two-layer LSTM, two-layer GRU, LSTM–GRU, and GRU–LSTM were effective for this application. The two-layer LSTM performed a 64.4%, 49.3%, and 53.3% improvement in attitude errors, the two-layer GRU performed a 56.3%, 49.3%, and 47.9% decrease in attitude errors, and the attitude errors of LSTM–GRU decreased by 72.2%, 69.3%, and 58.4%;

(2) With a limited training dataset, LSTM–GRU outperformed GRU–LSTM; LSTM–GRU had a large training loss, but the LSTM–GRU still showed an improvement compared with the two-layer LSTM and two-layer GRU.

Future works might include: It might be meaningful to explore the LSTM and GRU with more layers, which might give better performance; a dynamic trajectory could be employed to evaluate the performance of the artificial intelligence in this application; as this paper deals only with MRMS gyroscope de-noising, artificial intelligence could be integrated with the GNSS/INS method to improve the accuracy during GNSS signal outages.

Author Contributions: C.J. proposed the idea, and written the first version of this paper. S.C. revised the paper, discussed the paper and provided the funding for this paper. Y.C. revised the paper and guided the paper writing. Y.B. was the supervisor of C.J., and reviewed the paper before being submitted. W.L. given valuable advice on the paper writing and revision. W.T. and J.G. helped to collect and process the data.

Funding: This research was funded by the Fundamental Research Funds for the Central Universities (Grant No. 30917011105); the National Defense Basic Scientific Research program of China (Grant No. JCKY2016606B004); special grade of the financial support from the China Postdoctoral Science Foundation with grant number (Grant No. 2016T90461).

Acknowledgments: The author gratefully acknowledges the financial support from China Scholarship Council (CSC, Grant No. 201706840087). Thanks for the data collecting and processing support from Lin Han, Boya Zhang and Longjiang Fan. If you want the source code and data, please e-mail me, and I will share the all the materials.

Conflicts of Interest: The authors declare no conflict of interest.

References

1. Dow, J.M.; Ruth, E.N.; Chris, R. The international GNSS service in a changing landscape of global navigation satellite systems. *J. Geod.* **2009**, *83*, 191–198. [CrossRef]
2. Montenbruck, O.; Steigenberger, P.; Khachikyan, R.; Weber, G.; Langley, R.B.; Mervart, L.; Hugentobler, U. IGS-MGEX: Preparing the ground for multi-constellation GNSS science. *Inside GNSS* **2014**, *9*, 42–49.
3. Hewitson, S.; Wang, J. GNSS receiver autonomous integrity monitoring (RAIM) performance analysis. *GPS Solut.* **2006**, *10*, 155–170. [CrossRef]
4. Lashley, M.; Bevly, D.M.; Hung, J.Y. Performance analysis of vector tracking algorithms for weak GPS signals in high dynamics. *IEEE J. Sel. Top. Sign. Process.* **2009**, *3*, 661–673. [CrossRef]
5. Jiang, C.; Chen, S.; Chen, Y.; Bo, Y.; Wang, C.; Tao, W. Performance analysis of GNSS Vector tracking loop based GNSS/CSAC integrated navigation system. *J. Aeronaut. Astronaut. Aviat.* **2017**, *49*, 289–297.

6.   Chen, R.; Chen, Y.; Pei, L.; Chen, W.; Liu, J.; Kuusniemi, H.; Takala, J. A DSP-based multi-sensor multi-network positioning platform. In Proceedings of the 22nd International Technical Meeting of The Satellite Division of the Institute of Navigation, Savannah, GA, USA, 22–25 September 2009; pp. 615–621.

7.   Chen, Y.; Tang, J.; Jiang, C.; Zhu, L.; Lehtomäki, M.; Kaartinen, H.; Zhou, H. The accuracy comparison of three simultaneous localization and mapping (SLAM)-based indoor mapping technologies. *Sensors* **2018**, *18*, 3228. [CrossRef] [PubMed]

8.   Jiang, C.; Chen, S.; Chen, Y.; Zhang, B.; Feng, Z.; Zhou, H.; Bo, Y. A MEMS IMU de-noising method using long short term memory recurrent neural networks (LSTM-RNN). *Sensors* **2018**, *18*, 3470. [CrossRef]

9.   Chiang, K.W.; Huang, Y.W. An intelligent navigator for seamless INS/GPS integrated land vehicle navigation applications. *Appl. Soft Comput.* **2008**, *8*, 722–733. [CrossRef]

10.   Chiang, K.W.; Duong, T.T.; Liao, J.K. The performance analysis of a real-time integrated INS/GPS vehicle navigation system with abnormal GPS measurement elimination. *Sensors* **2013**, *13*, 10599–10622. [CrossRef]

11.   Syed, Z.F.; Aggarwal, P.; Goodall, C.; Niu, X.; El-Sheimy, N. A new multi-position calibration method for MEMS inertial navigation systems. *Meas. Sci. Technol.* **2007**, *18*, 1897. [CrossRef]

12.   Cho, S.Y.; Park, C.G. MEMS based pedestrian navigation system. *J. Navig.* **2006**, *59*, 135–153. [CrossRef]

13.   Brown, A.K. GPS/INS uses low-cost MEMS IMU. *IEEE Aerosp. Electron. Syst. Mag.* **2005**, *20*, 3–10. [CrossRef]

14.   Jiang, C.H.; Chen, S.; Chen, Y.Y.; Bo, Y.M. Research on chip scale atomic clock driven GNSS/SINS deeply coupled navigation system for augmented performance. *IET Radar. Sonar Navig.* **2018**. [CrossRef]

15.   Ning, Y.; Wang, J.; Han, H.; Tan, X.; Liu, T. an optimal radial basis function neural network enhanced adaptive robust Kalman filter for GNSS/INS integrated systems in complex urban areas. *Sensors* **2018**, *18*, 3091. [CrossRef] [PubMed]

16.   Niu, X.; Nassar, S.; El-Sheimy, N. An accurate land-vehicle MEMS IMU/GPS navigation system using 3D auxiliary velocity updates. *Navigation* **2007**, *54*, 177–188. [CrossRef]

17.   Li, W.; Wang, J. Effective adaptive Kalman filter for MEMS-IMU/magnetometers integrated attitude and heading reference systems. *J. Navig.* **2013**, *66*, 99–113. [CrossRef]

18.   Bhatt, D.; Aggarwal, P.; Devabhaktuni, V.; Bhattacharya, P. A novel hybrid fusion algorithm to bridge the period of GPS outages using low-cost INS. *Expert Syst. Appl.* **2014**, *41*, 2166–2173. [CrossRef]

19.   El-Sheimy, N.; Hou, H.Y.; Niu, X.J. Analysis and modeling of inertial sensors using Allan variance. *IEEE Trans. Instrum. Meas.* **2008**, *57*, 140–149. [CrossRef]

20.   Allan, D.W. Historicity, strengths, and weaknesses of Allan variances and their general applications. *Gyroscopy Navig.* **2016**, *7*, 1–17. [CrossRef]

21.   Radi, A.; Nassar, S.; El-Sheimy, N. Stochastic error modeling of smartphone inertial sensors for navigation in varying dynamic conditions. *Gyroscopy Navig.* **2018**, *9*, 76–95. [CrossRef]

22.   Aggarwal, P.; Syed, Z.; Niu, X.X.; El-Sheimy, N. A standard testing and calibration procedure for low cost MEMS inertial sensors and units. *J. Navig.* **2008**, *61*, 323–336. [CrossRef]

23.   Wang, D.; Dong, Y.; Li, Q.; Li, Z.; Wu, J. Using Allan variance to improve stochastic modeling for accurate GNSS/INS integrated navigation. *GPS Solut.* **2018**, *22*, 53. [CrossRef]

24.   Zhang, Q.; Wang, X.; Wang, S.; Pei, C. Application of improved fast dynamic Allan variance for the characterization of MEMS gyroscope on UAV. *J. Sens.* **2018**. [CrossRef]

25.   Wang, L.; Zhang, C.; Gao, S.; Wang, T.; Lin, T.; Li, X. Application of fast dynamic Allan variance for the characterization of FOGs-Based measurement while drilling. *Sensors* **2016**, *16*, 2078. [CrossRef] [PubMed]

26.   Su, W.P.; Hao, Y.S.; Li, Q.C. Arma-akf model of mems gyro rotation data random drift compensation. *Appl. Mech. Mater.* **2013**, *321–324*, 549–552. [CrossRef]

27.   Huang, L. Auto regressive moving average (ARMA) modeling method for Gyro random noise using a robust Kalman filter. *Sensors* **2015**, *15*, 25277–25286. [CrossRef] [PubMed]

28.   Khashei, M.; Bijari, M. A novel hybridization of artificial neural networks and ARIMA models for time series forecasting. *Appl. Soft Comput.* **2011**, *11*, 2664–2675. [CrossRef]

29.   Waegli, A.; Skaloud, J.; Guerrier, S.; Parés, M.E.; Colomina, I. Noise reduction and estimation in multiple micro-electro-mechanical inertial systems. *Meas. Sci. Technol.* **2010**, *21*, 156–158. [CrossRef]

30.   Bhatt, D.; Priyanka, A.; Prabir, B.; Vijay, D. An enhanced mems error modeling approach based on nu-support vector regression. *Sensors* **2012**, *12*, 9448–9466. [CrossRef]

31. Xing, H.F.; Hou, Bo.; Lin, Z.H.; Guo, M.F. Modeling and compensation of random drift of MEMS gyroscopes based on least squares support vector machine optimized by chaotic particle swarm optimization. *Sensors* **2017**, *17*, 2335. [CrossRef]

32. Jiang, C.; Chen, S.; Chen, Y.; Bo, Y.; Han, L.; Guo, J.; Feng, Z.; Zhou, H. Performance Analysis of a Deep Simple Recurrent Unit Recurrent Neural Network (SRU-RNN) in MEMS Gyroscope De-Noising. *Sensors* **2018**, *18*, 4471. [CrossRef]

33. Understanding LSTM Networks. Available online: https://colah.github.io/posts/20150--8-Understanding-LSTMs/ (accessed on 27 August 2015).

34. Hosseinyalamdary, S. Deep Kalman filter: Simultaneous multi-sensor integration and modelling; A GNSS/IMU case study. *Sensors* **2018**, *18*, 1316. [CrossRef] [PubMed]

35. Chung, J.; Gulcehre, C.; Cho, K.; Bengio, Y. Gated feedback recurrent neural networks. In Proceedings of the International Conference on Machine Learning, Lille, France, 6–11 July 2015; pp. 2067–2075.

36. Gers, F.A.; Schraudolph, N.N.; Schmidhuber, J. Learning precise timing with LSTM recurrent networks. *J. Mach. Learn. Res.* **2002**, *3*, 115–143.

37. Ordóñez, F.J.; Roggen, D. Deep convolutional and LSTM recurrent neural networks for multimodal wearable activity recognition. *Sensors* **2016**, *16*, 115. [CrossRef] [PubMed]

38. MSI3200. Available online: http://www.mtmems.com/product_view.asp?id=28 (accessed on 5 August 2017).

![electronics logo] *electronics*

MDPI

Article

Operational Improvement of Interior Permanent Magnet Synchronous Motor Using Fuzzy Field-Weakening Control

Ming-Shyan Wang [1],*, Min-Fu Hsieh [2] and Hsin-Yu Lin [1]

[1]   Department of Electrical Engineering, Southern Taiwan University of Science and Technology, 1, Nan-Tai St., Yung Kang District, Tainan City 710, Taiwan; ma420203@stust.edu.tw
[2]   Department of Electrical Engineering, National Cheng Kung University, 1, University Road, East District, Tainan City 701, Taiwan; mfhsieh@mail.ncku.edu.tw
*   Correspondence: mswang@stust.edu.tw; Tel.: + 886-6-2533131#3328

Received: 27 September 2018; Accepted: 17 December 2018; Published: 18 December 2018

Abstract: This paper considers the fuzzy control design of maximum torque per ampere (MTPA) and maximum torque per voltage (MTPV) for the interior permanent magnet synchronous motor (IPMSM) control system that is capable of reducing computation burden, improving torque output, and widening the speed range. In the entire motor speed range, three control methods, i.e., the MTPA, flux weakening, and MTPV methods may be applied depending on current and voltage statuses. The simulation using MATLAB/Simulink is first conducted and then in order to speed up the development, hardware-in-the-loop (HIL) is adopted to verify the effectiveness of the proposed fuzzy MTPA and MTPV control for the IPMSM system.

Keywords: interior permanent magnet synchronous motor; fuzzy logic; maximum torque per ampere (MTPA); field weakening; maximum torque per voltage (MTPV); hardware in the loop (HIL)

---

1. Introduction

The consumption of energy by industrial and domestic electric motors per year occupies 46.2% of the global electrical demand [1]. In addition, electrical vehicles (EVs) have continuously attracted the attention of researchers and companies [2,3]. Due to its high power density, efficiency, low maintenance cost, and wide range speed regulation, an interior permanent magnet synchronous motor (IPMSM) is an attractive selection for EVs and hybrid electric vehicles (HEVs) [4]. In order to achieve high compactness and to abstain from a multi-stage gearbox, an electrical traction drive usually has a wide constant power speed range (up to 12,000 rpm) [5] to cover the high driving speed of the EVs.

Traditionally, the d-axis stator current component of the IPMSM is set at zero in the current/torque control for easier design. However, the efficiency of the drive system cannot be optimized without controlling the air gap flux. As a result, the reluctance torque of the IPMSM, which is one of its advantages as compared with a surface-mounted PMSM, cannot be utilized. In order to improve this and extend the speed range, it is necessary to maximize the IPMSM torque output appropriately along the optimal current trajectory over the whole speed range. The typical IPMSM current trajectory on the $i_d - i_q$ plane may consist of three or four paths or regions [6–8], as shown in Figure 1. Region 1 (curve OP) is the maximum torque per ampere (MTPA) operation that generates the required torque with a minimum phase current. Region 2 (hyperbola PQ) moves the current trajectory away from the MTPA curve along the torque hyperbola. Optional Region 3 (arc QB) starts the flux-weakening operation and moves the current trajectory along the current limitation circle. Region 4 (curve BE) is the maximum torque per voltage (MTPV) operation to generate possible maximum torque under the inverter voltage limitation.

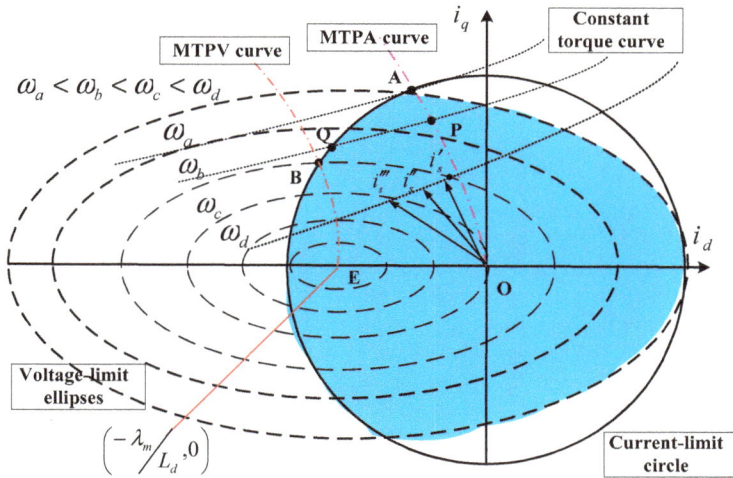

Figure 1. Curves under the drive current and voltage constraints.

The MTPA control means that the ratio between the produced torque and the current amplitude is maximized through properly selecting the current space vector as a torque function. The MTPV control is used to increase the motor speed further and extend the torque control capability under the voltage-limited maximum output current vector control if the center of the voltage-limit ellipses of the motor lies inside the current-limit circle. Generally, this is easy by using the optimum current space vector to understand and conduct the mathematical and graphical analyses of the constant-torque and constant-current loci. However, it requires precise values of motor parameters, such as direct- and quadrature-axis inductances and the flux linkages, and the stator resistance. Essentially, those are not easy to be found over a wide speed range just by considering an optimization problem that considers torque error minimization as the objective function with the inverter voltage and current as constraints. Furthermore, it is usually time-consuming and difficult to directly solve this problem from the closed-form solutions of nonlinear torque, voltage, and current equations.

The recursive least square (RLS) estimation is proposed in Reference [5] to identify the variable parameters of motor characteristics online for a high speed and high performance IPMSM. It leads to an improvement of the efficiency by around 6% from 79% using the method with constant parameters. A model linearization-based approach is proposed in Reference [6]. The optimal torque control problem over a wide speed range is divided into two sub-optimizations to solve them sequentially to simplify the calculation. However, only simulation results were provided. Linear field-weakening control (LFC) of IPMSM is proposed in Reference [7] to prevent losing current control due to the saturation of current regulators for nonlinear field-weakening in wide speed range applications. As a result, the operation range is only divided into three areas, that is, constant torque linear area, constant power transition linear area, and constant direct axis current area. However, due to the uncertainties of the characteristics of the voltage source, inductance, and the permanent magnet flux linkage, the torque and power performance needs to be improved. Direct torque and flux control (DTFC) is used to especially investigate the deep flux-weakening control of an IPMSM along the maximum torque per voltage (MTPV) trajectory in the torque-flux plane [8,9]. A feedforward look-up-table-based method is proposed [10] to consider the motor parameter variation as the motor operates along the MTPV region by controlling both the voltage and current vectors. In addition to a look-up table, this method was relatively complicated [11].

It is well known that intelligent control has the advantages of being without need of an exact system dynamic model, simplicity, less intensive mathematical design, and is suitable for dealing with the nonlinearities and uncertainties. Fuzzy logic control (FLC), neural network control (NNC), and neurofuzzy control (NFC), etc., belong to intelligent control. In addition, FLC is the simplest one for implementation. Therefore, a fuzzy logic controller is applied in many fields [12–19]. Contrary to the conventional FLC of a IPMSM drive with zero d-axis current, a simplified fuzzy speed controller with MTPA incorporated for the IPMSM drive is proposed [20]. It simplified the d-axis current around some operating point to get the electromagnetic torque. The authors in Reference [21] proposed the FLC based IPMSM drive with variable d-axis and q-axis current equations to investigate the performance and compare it to that obtained from the drive via calculating dynamic MTPA equations using MATLAB/Simulink. An online loss-minimization MTPA algorithm is further integrated with a FLC-based IPMSM drive to yield high efficiency and high dynamic performance over a wide speed range [22]. A fuzzy control algorithm based on the operation time of zero space voltage vectors [23] to derive d-axis current for the field-weakening control of PMSM is proposed. The FLC, which outputs d- and q-axis currents of IPMSM, is proposed based on conventional MTPA operation using simplified equation [24] and the field-weakening operation while maintaining current and voltage constraints. The simulation results of fuzzy MTPA and MTPV control of IPMSM using MATLAB/Simulink is shown to verify the proposed algorithm [13]. Based on the following reasons, a simple intelligent control algorithm, less computation time, easy implementation by microcontroller/digital signal processor, many successful examples, and MTPA, field weakening, and MTPV all designed over the full speed span in this paper, FLC was adopted.

It usually takes a long time and costs much to complete a new designed IPMSM and to verify its performance. A hardware-in-the-loop (HIL) system will be a workable substitute. The MR2 is one HIL system [25]. It is a virtual reality integration platform designed for controller development, verification, and testing. It enables research and development personnel to conduct product development, verification, and debugging in a safe and convenient real-time simulation environment. Connecting the controller's I/O terminals, MR2 can quickly receive the signals from and return the operating results in a very short period time to the controller. Users may choose motors according to their needs. Five kinds of system parameters are included: input voltage and feedback signal scaling setup, grid and rectifier module setup, motor parameters setup, velocity sensor module and load torque setup, and customized analog output signals setup.

In this paper, the IPMSM model and the concept of MTPA and MTPV are introduced in Section 2. In Section 3, the fuzzy logic control system (FLC) is described. Simulations using MATLAB/Simulink (The MathWorks Inc., 1 Apple Hill Drive Natick, MA, USA) and experiments using the HIL system of the proposed IPMSM drive are shown in Section 4. Finally, conclusions are given in Section 5.

2. Modeling IPMSM, MTPA, and MTPV

The voltage equations of the IPMSM in the dq-frame are expressed using Equations (1)–(4) [4–7],

$$v_d = R_s i_d + p\lambda_d - \omega_e \lambda_q \tag{1}$$

$$v_q = R_s i_q + p\lambda_d - \omega_e \lambda_d \tag{2}$$

$$\lambda_d = L_d i_d + \lambda_m \tag{3}$$

$$\lambda_q = L_q i_q \tag{4}$$

where v_d and v_q, i_d and i_q, λ_d and λ_q, and L_d and L_q denote the d- and q-axis voltages, currents, flux linkages, and inductances, respectively; R_s is the phase resistance; p is the differential operator; ω_e is the electric speed; and λ_m is the magnet flux linkage. Combining Equations (1) to (4), we obtain:

$$v_d = R_s i_d + L_d p i_d - \omega_e L_q i_q \tag{5}$$

$$v_q = R_s i_q + L_q p i_q + \omega_e(\lambda_m L_q i_q) \tag{6}$$

The electromagnetic torque is then expressed using Equation (7):

$$T_e = \frac{3}{2}\frac{P}{2}\left(\lambda_d i_q - \lambda_q i_d\right) = \frac{3}{2}\frac{P}{2}\{\lambda_m i_q + (L_d - L_q)i_q i_d\} \tag{7}$$

where P is the number of poles, $\frac{3}{2}\frac{P}{2}\lambda_m i_q$ stands for the magnetic torque, and $\frac{3}{2}\frac{P}{2}(L_d - L_q)i_q i_d$ is the reluctance torque.

In the real case, the current and voltage are subject to real constraints:

$$i_q^2 + i_d^2 \le I_{s\,max}^2 \tag{8}$$

$$v_q^2 + v_d^2 \le V_{s\,max}^2 \tag{9}$$

where $I_{s\,max}$ is the rated current of the motor and $V_{s\,max}$ is the maximum voltage dependent on the dc-link bus and pulse-width modulation (PWM) method. A bold circle and its interior in Figure 1 depict Equation (8). For simplicity of analysis, at a steady state of the motor running, the voltage drop on resistance was neglected from Equations (5) and (6):

$$V_d = -\omega_e L_q i_q \tag{10}$$

$$V_q = \omega_e L_d i_d + \omega_e \lambda_m \tag{11}$$

Substituting Equations (10) and (11) into Equation (9) and rearranging it, we obtain ellipses and their interiors with different speeds as shown in Figure 1 given by Equation (12):

$$\frac{\left(i_d + \frac{\lambda_m}{L_d}\right)^2}{\left(\frac{V_{s\,max}}{L_d\omega_e}\right)^2} + \frac{i_q^2}{\left(\frac{V_{s\,max}}{L_q\omega_e}\right)^2} \le 1 \tag{12}$$

where $(-\lambda_m/L_d, 0)$ is the center; $V_{s\,max}/(\omega_e L_d)$ and $V_{s\,max}/(\omega_e L_q)$ are the lengths of the semi major and the semi minor axes, respectively, for the dashed ellipses; and electric speeds of $\omega_a < \omega_b < \omega_c < \omega_d$. The overlapped area of the circle and ellipses denotes the operable region of the motor drive system. It is easy to find that the lengths of the semi major and the semi minor axes, as well as the overlapped area, will shrink as the motor speed increases. At the same time, the loci of available current vectors are progressively reduced. This is due to the increase of the back electromotive force (EMF) amplitude. That is to say, the voltage limit plays an increasingly dominant role at higher speed operation. The point of tangency using the circle and the ellipse denotes the speed at no load under a maximum operable voltage, where the motor does not output torque. The operable region of the motor system includes four quadrants. The motor mode occupies the first and second quadrants and the generator mode operates in the third and fourth quadrants.

The main purpose of motor torque at low speed is to accelerate the drive system. As the d-axis current is held at zero, the voltage limit ellipse will force the q-axis current to decrease rapidly when the motor speed approaches and exceeds the rated speed. As a result, a fast torque drop occurs. Controlling the phase between the back EMF and the current vector will result in the effect of flux weakening, that is, a negative d-axis current is designed. Under the current constraint of Equation (10), many stator currents can be set to satisfy some specified torque requirement. For example, various stator current vectors—i_s', i_s'', and i_s''' in Figure 1—provide the same torque under different motor speeds. As a result, the torque–speed envelope is considerably expanded as compared with the case of a zero d-axis current. Furthermore, it will benefit the system by using the optimal torque control to reduce power dissipation or raise efficiency via supplying the same torque using less current.

The relationship between the d- and q-axis stator currents is shown in Figure 2 with equations as follows:

$$i_s = i_d + j i_q$$

$$i_d = i_s \cos \beta$$

$$i_q = i_s \sin \beta$$

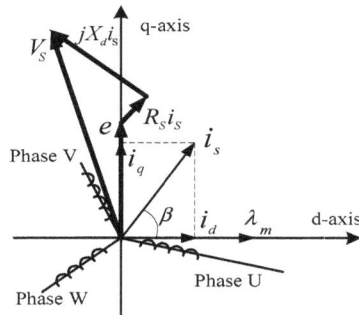

Figure 2. Voltage and current vectors.

A new torque equation is generated by substituting Equation (13) into Equation (7):

$$T_e = \frac{3}{2}\frac{P}{2}\left[\lambda_m i_s \sin \beta - \frac{1}{2}\left(L_q - L_d\right)i_s^2 \sin 2\beta\right] \tag{13}$$

The maximum torque per ampere (MTPA) is obtained by differentiating Equation (14) with respect to β and setting it to be zero. The angle for the maximum torque output is:

$$\beta_{max} = cos^{-1}\left[\frac{\lambda_m - \sqrt{\lambda_m + 8\left(L_q - L_d\right)^2 i_s^2}}{4\left(L_q - L_d\right)i_s}\right] \tag{14}$$

and the d-axis current for MTPA is:

$$i_{dPA} = \frac{\lambda_m - \sqrt{\lambda_m^2 + 8\left(L_q - L_d\right)^2 i_q^2}}{2\left(L_q - L_d\right)} \tag{15}$$

Using Equation (16), the curve of the MTPA shown in Figure 1 is found in the field weakening region based on various loads. The intersections of the MTPA curve, current circle, and voltage ellipses denote the maximum output torque of the motor running at different speeds. Point A in Figure 1 depicts the rated torque and speed of the motor. The d-axis current at point A will be:

$$i_{dA} = \frac{\lambda_m - \sqrt{\lambda_m^2 + 8\left(L_d - L_q\right)^2 i_{s\,max}^2}}{4\left(L_q - L_d\right)} \tag{16}$$

In the MTPA trajectory, the d- and q-axis current components of the current space vector are found from the intersection between the constant-torque hyperbola and the constant-current circle, with the constraint of minimum length of the current space vector (i.e., the constant-current locus is tangent to the constant-torque locus).

For wider operational speed in the constant power region, a technique called the maximum torque per voltage (MTPV) control is considered by controlling the current vector such that the torque per flux linkage becomes maximal. In the flux-weakening range, if the characteristic current $|\lambda_m/L_d|$ is less than the rated motor current, the torque controllability can be extended by using MTPV control [9]. The expressions of i_d and i_q for MTPV can be given as:

$$i_d = -\frac{\lambda_m - \lambda_d}{L_d} \tag{17a}$$

$$i_q = -\frac{\sqrt{\lambda_s{}^2 - \lambda_d{}^2}}{L_q} \tag{17b}$$

From Equation (11), Equation (9) and the stator flux are given as:

$$V_{s\,max} \geq \omega_e{}^2 \left(L_q i_q\right)^2 + \omega_e{}^2 (L_d i_d + \lambda_m)^2 \tag{18}$$

$$\lambda_s = \sqrt{\lambda_d{}^2 + \lambda_q{}^2} = \frac{V_{s\,max}}{\omega_e} \tag{19}$$

As a result, the new equation from Equation (7) for torque is:

$$T_e = \left(\frac{3P}{4}\right)\left(\lambda_d\left(\frac{\sqrt{\lambda_s{}^2 - \lambda_d{}^2}}{L_q}\right) + \sqrt{\lambda_s{}^2 - \lambda_d{}^2}\left(\frac{\lambda_m - \lambda_d}{L_d}\right)\right) \tag{20}$$

Similarly, if we differentiate Equation (21) with respect to λ_d and set it to be zero, that is, $\frac{\partial T_e}{\partial \lambda_d} = 0$, we obtain the d-axis flux and current using the following equations for MTPV:

$$\lambda_{d,max} = \frac{-L_q\lambda_m + \sqrt{\left(L_q\lambda_m\right)^2 + 8\left(L_d - L_q\right)^2\left(\frac{V_{s\,max}}{\omega_e}\right)^2}}{4\left(L_d - L_q\right)} \tag{21}$$

$$i_{d,max} = -\frac{\lambda_m - \lambda_{d,max}}{L_d} \tag{22}$$

The MTPV curve is shown in Figure 1.

In summary, to produce the maximum output power in all speed ranges considering the condition of both the current and the voltage limits, the optimum current vector is chosen as follows.

- Region I ($\omega_e < \omega_{eA}$): i_d and i_q are constant values given using Equation (13). The current vector is fixed at point A in Figure 1. In this region, $i_s = I_{s\,max}$ and $V_s = \sqrt{v_d^2 + v_q^2} < V_{s\,max}$.
- Region II ($\omega_{eA} < \omega_e < \omega_{eB}$): i_d and i_q are chosen as the intersection of the current-limit circle and the voltage-limit ellipses. The current vector moves from point A to B along the current-limit circle as the motor speed increases. In this region, $i_s = I_{s\,max}$ and $V_s = V_{s\,max}$.
- Region III ($\omega_e > \omega_{eB}$): i_d and i_q are given using Equation (18). The current vector moves from point B to the center of the ellipse along the voltage-limit maximum-output trajectory. In this region, $i_s < I_{s\,max}$ and $V_s = V_{s\,max}$.

3. Fuzzy Logic Control

Figure 3 shows the basic blocks of a fuzzy logic control (FLC) system, input variables, the knowledge base (data and rule bases), the inference engine, the fuzzification interface, the defuzzification interface, and output variables. The input and output variables are crisp. The fuzzification interface converts the crisp inputs to fuzzy sets and the defuzzification interface converts these fuzzy conclusions back into the crisp outputs to ensure the desired performance. The fuzzy controller is essentially an artificial and real-time decision-maker in a closed-loop system based on the experts' experience.

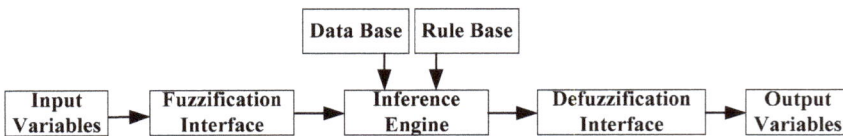

Figure 3. Basic architecture of a fuzzy control system.

A fuzzy logic control system is used to optimize the MTPA and MTPV as shown in Figure 4a,b, respectively. The linguistic values Z, S, M, B, and VB represent zero, small, medium, big, and very big, respectively. The membership functions of these input and output fuzzy variables of fuzzy MTPA with normalization are shown in Figure 5, in which the triangular form of the membership functions eases calculating to reduce computation burden of fuzzy MTPA and MTPV. The design methodology does not focus on the specific motor or operation point. It is applicable to any IPMSM. There are 25 rules given in Figure 6. The first rule is shown as follows.

Rule 1: If ω_e is Z and i_q is Z, then i_d^* is S.

The rule numbers are listed in the parentheses of the table. Rules 13, 14, 18, and 19 are more often triggered. Equation (24) describes the min-min-max inference and mean of height method in the FLC system:

$$\hat{y} = \frac{\sum_{i=1}^{n} f_i \times g_i}{\sum_{i=1}^{n} f_i} \tag{23}$$

where g_i is the center of the *i*th fuzzy set and f_i is its height, and $\hat{y}$ is the center average.

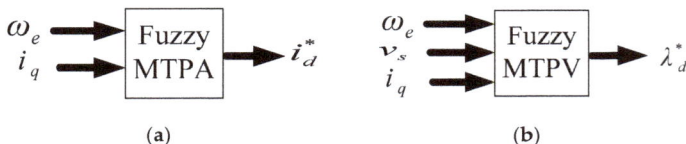

Figure 4. Fuzzy control based MTPA (**a**), and MTPV (**b**).

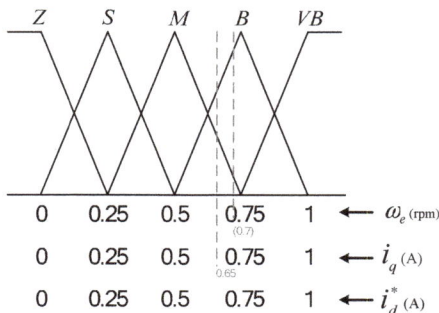

Figure 5. Membership functions of fuzzy MTPA.

Similarly, the 25 rules for fuzzy MTPV are listed in Figure 7 since the input variable i_q is only used to check if the magnitude of the stator current vector is larger than the limited value. That is: if $\sqrt{i_q^2 + i_d^2} > I_{s\,max}$, then $i_q = \sqrt{I_{s\,max}^2 - i_d^2}$; else, Rule 1: if ω_e is Z and v_s is Z, then λ_d^* is S. Rules 8, 9, 13, and 14 are triggered more often.

ω_e \\ i_q	Z	S	M	B	VB
Z	S(1)	M(6)	B(11)	B(16)	VB(21)
S	S(2)	M(7)	M(12)	VB(17)	VB(22)
M	S(3)	S(8)	M(13)	B(18)	VB(23)
B	Z(4)	S(9)	M(14)	B(19)	VB(24)
VB	Z(5)	S(10)	M(15)	B(20)	B(25)

Figure 6. MTPA rule table.

ω_e \\ v_s	Z	S	M	B	VB
Z	S(1)	M(6)	B(11)	B(16)	VB(21)
S	S(2)	M(7)	M(12)	VB(17)	VB(22)
M	S(3)	S(8)	M(13)	B(18)	VB(23)
B	Z(4)	S(9)	M(14)	B(19)	VB(24)
VB	Z(5)	S(10)	M(15)	B(20)	B(25)

Figure 7. MTPV rule table.

4. Simulation and Experimental Results

The block diagram of the proposed control system for simulation and experiment, including the test IPMSM, its drive, and servo motor for loading, is shown in Figure 8. Each block is modeled according to its transfer function. The parameters of IPMSM are listed in Figure 9 (taken from HIL). There are three proportional and integral (PI) controllers used in the speed and current control loops, respectively. Based on References [26–29], the rule of thumb [30], and our experience, the proportional and integral gains under per-unit processing of speed control loop, d-axis current loop, and q-axis current loop were 0.01 and 0.1, 0.1 and 0.01, and 0.1 and 0.001, respectively. These parameters were adopted and the same in the simulations and experiments.

By using MATLAB/Simulink, Figures 10 and 11 show the simulation results of curves of load versus current, respectively, at 300 rpm and 2000 rpm using a fixed torque angle method and fuzzy MTPA control with the loads of 1, 2, 3, 4, and 5 Nm. Under the same load, a lower stator current was needed for fuzzy MTPA control. The difference in stator current for both control methods was larger at a heavy load and for a lower speed range. Figure 12 shows the simulation results of the torque–speed curves (called T–N curves hereafter) using a fixed torque angle method and fuzzy MTPA/MTPV control with the load of 5 Nm, double that of the rated torque. The constant torque range shrunk to 1000 rpm, half of the rated speed of 2000 rpm, but the T–N curve using fuzzy MTPA/MTPV control provided a larger torque output and a larger speed range in the constant power region than those using a fixed torque angle method.

MR2 HIL and the related circuits for experiments are shown in Figure 13. Figures 14 and 15 show the results of curves of load versus current at 900 rpm and 2000 rpm using a fixed torque angle method and fuzzy MTPA control with the loads of 1, 2, 3, 4, and 5 Nm. The results provided the same conclusions of simulation, but with a larger current at light load, about 20 A. In addition, the differences of stator current for both control was almost independent of motor speed. Figure 16 depicts the locus of stator current (bold red) with a load of 3 Nm under the drive current and voltage constraints. The corresponding T–N curve using MTPA, field weakening, and MTPV control in the whole speed range is shown in Figure 17. It can be found that the constant torque region ended at the speed of 1900 rpm, but it had a larger torque output in the constant power region than that of the rated power. For the situation of rated load, the locus of the stator current (bold red) is drawn in Figure 18. Figure 19 shows T–N curve initially under MTPA control, then under field weakening control from the rated speed of 2000 rpm, and finally under MTPV control from speed of 4000 rpm to the center of ellipses, about 5600 rpm. It was verified that the proposed algorithm not only precisely extended the wider speed range at the constant power region from the rated speed of 2000 rpm to 4000 rpm, but also prolonged the constant torque range and provided a larger torque output in the constant power region.

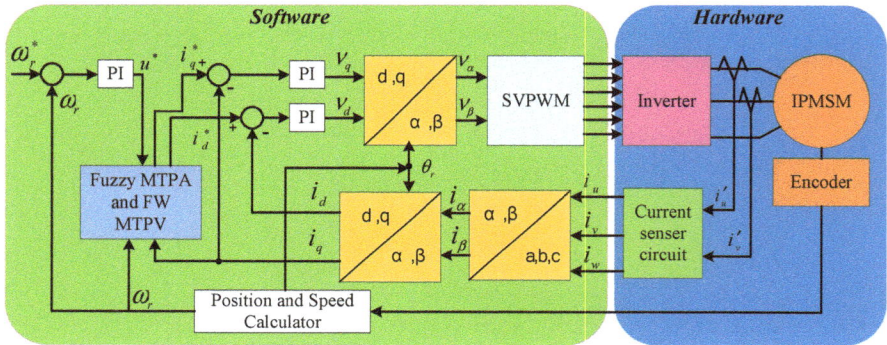

Figure 8. Block diagram of the fuzzy controlled system.

Pr. Names	Abbr.	Unit	Value
Coordinate alignment	dq	Pi	1
Stator Winding Resistance	Rs	Ohm	0.084848
d-axis Inductance	Ld	mH	0.13073
q-axis Inductance	Lq	mH	0.33208
Back EMF Source		pu.	0
Back EMF constant (L-L)	Ke	V/kRPM	4.402
Rated Torque	Te	Nm	2.488
Rated Current	Irated	A	40
Poles	P	pu.	4
dq-axis Inductance Source		pu.	0
Inductance File Name	Path	N/A	
System Inertia	Jm	Kg·m^2	0.01
System Damping	Bm	Kg·m^2/sec	0.002

Figure 9. Parameters of IPMSM (taken from HIL).

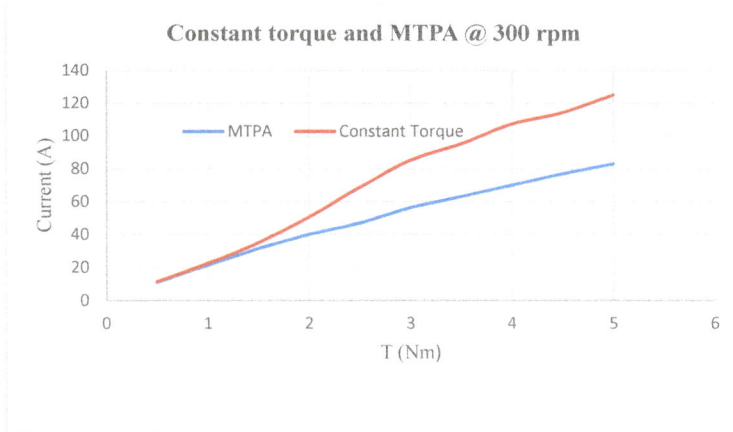

Figure 10. Curves of load vs current at 300 rpm by fixed torque angle method and fuzzy MTPA control.

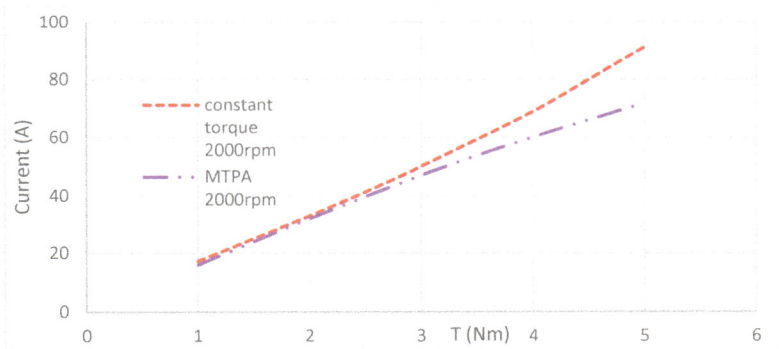

Figure 11. Curves of load vs current at 2000 rpm by fixed torque angle method and fuzzy MTPA control.

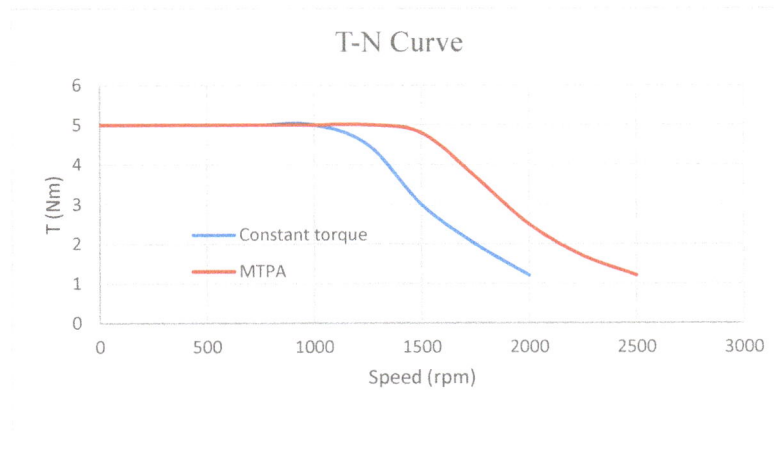

Figure 12. Torque–speed curves by fixed torque angle method and fuzzy MTPA/MTPV control.

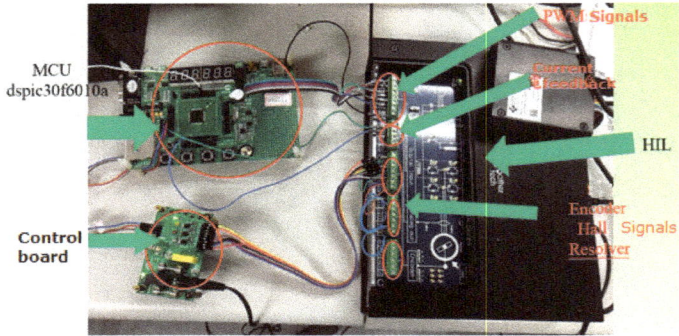

Figure 13. Hardware-in-the-loop (HIL) and the related circuits for experiments.

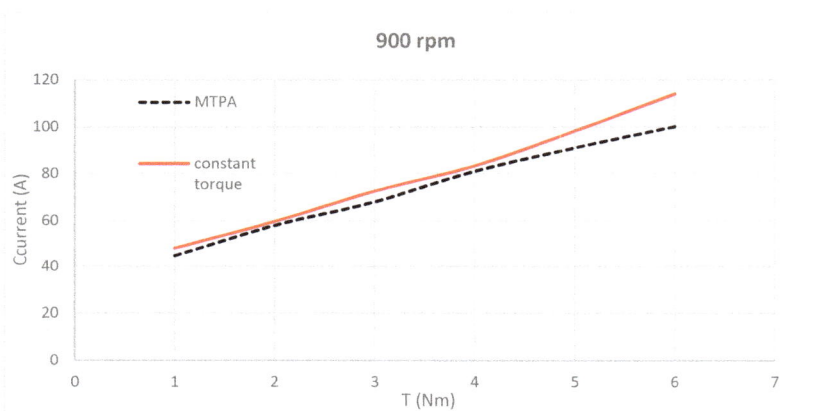

Figure 14. Curves of load vs current at 900 rpm using a fixed torque angle method and fuzzy MTPA control via HIL.

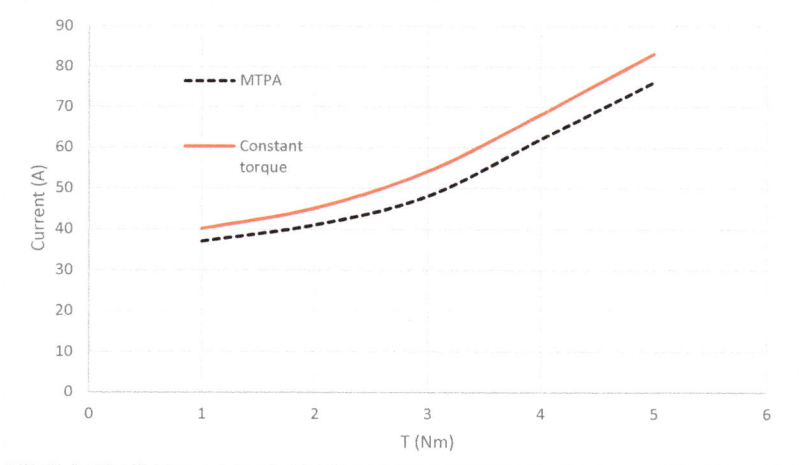

Figure 15. Curves of load vs current at 2000 rpm using a fixed torque angle and fuzzy MTPA control via HIL.

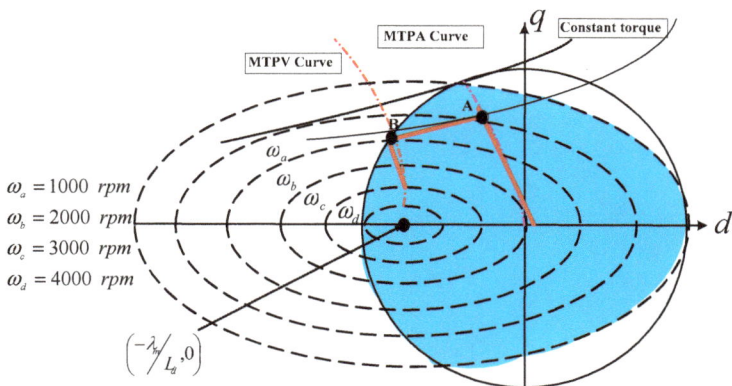

Figure 16. Locus of stator current (bold red) with load of 3 Nm under the drive current and voltage constraints.

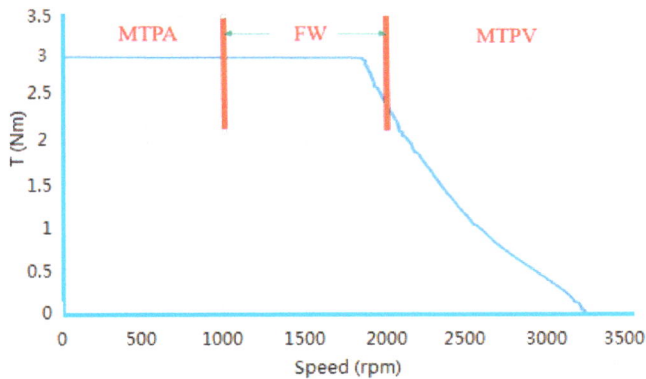

Figure 17. Torque–Speed Curve by MTPA and FW and MTPV control with load of 3 Nm in the whole speed range.

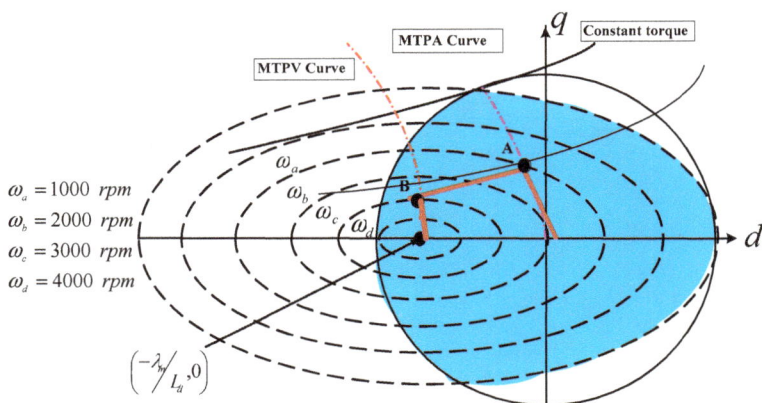

Figure 18. Locus of stator current (bold red) with load of 2.5 Nm under the drive current and voltage constraints.

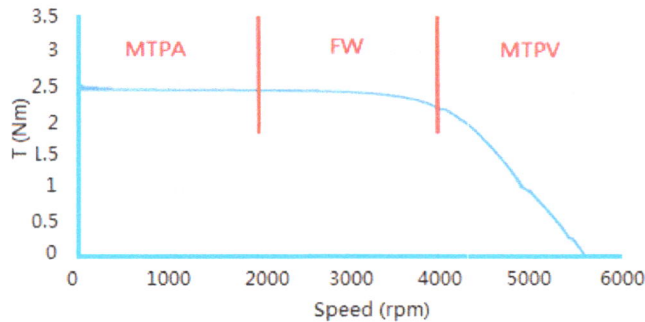

Figure 19. T-N Curve by MTPA and FW and MTPV control with load of 2.5 Nm in the whole speed range.

5. Conclusions

In this paper, the proposed fuzzy logic based controlled IPMSM drive system had extended the operating speed range and accompanied larger torque output using the maximum torque per ampere, field-weakening, and maximum torque per voltage techniques in various speed regions. In addition, the proposed system also mitigated the burden of the complex computation of torque control optimization problem. From the simulation results by MATLAB/Simulink shown in Figures 9–11, a lower stator current was needed for fuzzy MTPA control under the same load; the proposed system not only precisely extended wider speed range at the constant power region from the rated speed of 2000 rpm to 5000 rpm, but also prolonged the constant torque range and provided larger torque output in constant power region. For the experimental results using a MR2 HIL system depicted in Figures 13–15, the results provided the same conclusions of simulation; the proposed system initially took a load of 2.5 Nm under MTPA control, then under field weakening control from the rated speed of 2000 rpm; and finally under MTPV control from speed of 4000 rpm to the center of ellipses, about 5600 rpm. Those results verified the effectiveness of the proposed fuzzy MTPA/MTPV control. However, some future works are necessary for us, such as studying new MTPA and MTPV algorithms, utilizing full inverter voltage to improve torque output, and more complex and advanced fuzzy control algorithms, to keep improving research.

Author Contributions: M.-S.W. conceived and designed the experiments; M.-F.H. designed the motor; H.-Y.L. performed the experiments; M.-S.W. and H.-Y.L. analyzed the data; M.-S.W. and M.-F.H. contributed materials and analytical tools; M.-S.W. wrote the paper.

Funding: This research was funded by Ministry of Science and Technology, Taiwan under contract Nos. of MOST 107-2622-E-218-008 -CC2, 106-2221-E-218-002-, and 106-2622-8-006 -001-.

References

1. Energy-Efficiency Policy Opportunities for Electric Motor-Driven Systems. Available online: http://www.iea.org/publications/freepublications/publication/ee_for_electricsystems.pdf (accessed on 15 February 2017).
2. Xu, Q.; Sun, J.; Tian, D.; Wang, W.; Huang, J.; Cui, S. Analysis and Design of a Compound-Structure Permanent-Magnet Motor for Hybrid Electric Vehicles. *Energies* **2018**, *11*, 2156. [CrossRef]
3. Xu, Q.W.; Sun, J.; Su, Y.M.; Chen, W.D.; Huang, J.S.; Cui, S.M. Study on the magnetic coupling and decoupling algorithm of electrical variable transmission. In Proceedings of the ICSEE 2017, LSMS 2017: Intelligent Computing, Networked Control, and Their Engineering Applications, Nanjing, China, 22–24 September 2017; Springer: Singapore, 2017; Volume 762, pp. 146–155.

4. Takeda, M.Y.; Hirasa, T. Expansion of operating limits for permanent magnet motor by current vector control considering inverter capacity. *IEEE Trans. Ind. Appl.* **1990**, *26*, 866–871.

5. Nguyen, Q.K.; Petrich, M.; Jörg, R.-S. Implementation of the MTPA and MTPV control with online parameter identification for a high speed IPMSM used as traction drive. In Proceedings of the 2014 International Power Electronics Conference (IPEC-Hiroshima 2014—ECCE ASIA), Hiroshima, Japan, 18–21 May 2014; pp. 318–323.

6. Yang, L.; Gao, R.; Yu, W.; Husain, I. A geometrical linearization approach for salient-pole PMSM optimal voltage/current constrained control over whole speed range. In Proceedings of the 2017 IEEE Energy Conversion Congress and Exposition (ECCE), Cincinnati, OH, USA, 1–5 October 2017; pp. 350–356.

7. Gallegos-Lopez, G.; Gunawan, F.S.; Walters, J.E. Optimum torque control of permanent-magnet AC Machines in the field-weakened region. *IEEE Trans. Ind. Appl.* **2005**, *41*, 1020–1028. [CrossRef]

8. Chen, K.; Sun, Y.; Liu, B. Interior Permanent Magnet Synchronous Motor Linear Field-Weakening Control. *IEEE Trans. Energy Convers.* **2016**, *31*, 159–164. [CrossRef]

9. Ekanayake, S.; Dutta, R.; Rahman, M.F.; Xiao, D. Direct torque and flux control of interior permanent magnet synchronous machine in deep flux-weakening region. *IET Electr. Power Appl.* **2018**, *12*, 98–105. [CrossRef]

10. Bing, C.; Tesch, T.R. Torque Feedforward Control Technique for Permanent-Magnet Synchronous Motors. *IEEE Trans. Ind. Electron.* **2010**, *57*, 969–974. [CrossRef]

11. Ekanayake, S.; Dutta, R.; Rahman, M.F.; Xiao, D. Deep flux weakening control of a segmented interior permanent magnet synchronous motor with maximum torque per voltage control. In Proceedings of the IECON 2015—41st Annual Conference of the IEEE Industrial Electronics Society, Yokohama, Japan, 9–12 November 2015; pp. 4802–4807.

12. Mohamed, Y.A.R.I.; Lee, T.K. Adaptive self-tuning MTPA vector controller of IPMSM drive system. *IEEE Trans. Energy Convers.* **2006**, *21*, 636–644. [CrossRef]

13. Wang, M.-S.; Hsieh, M.-F.; Syamsiana, I.N.; Fang, W.-C. Fuzzy Maximum Torque per Ampere and Maximum Torque per Voltage Control of Interior Permanent Magnet Synchronous Motor Drive. *Sens. Mater.* **2017**, *29*, 461–472. [CrossRef]

14. Wang, M.-S.; Hsieh, M.-F.; Kung, Y.-S.; Lin, G.T. Maximum Torque per Ampere Control of IPMSM Drive by Fuzzy Logic. *Microsyst. Technol.* **2016**, *22*, 1–8. [CrossRef]

15. Baldwin, J.F.; Lawry, J. A new approach to learning linguistic control rules. *Int. J. Uncertain. Fuzziness Knowl. Based Syst.* **2000**, *8*, 21–44. [CrossRef]

16. Škrjanc, I.; Blažič, S.; Matko, D. Direct fuzzy model-reference adaptive control. *Int. J. Intell. Syst.* **2002**, *17*, 943–963. [CrossRef]

17. Precup, R.-E.; Preitl, S.; Korondi, P. Fuzzy controllers with maximum sensitivity for servosystems. *IEEE Trans. Ind. Electron.* **2007**, *54*, 1298–1310. [CrossRef]

18. Chatterjee, A.; Chatterjee, R.; Matsuno, F.; Endo, T. Augmented stable fuzzy control for flexible robotic arm using LMI approach and neuro-fuzzy state space modeling. *IEEE Trans. Ind. Electron.* **2008**, *55*, 1256–1270. [CrossRef]

19. Vrkalovic, S.; Teban, T.-A.; Borlea, I.-D. Stable Takagi-Sugeno fuzzy control designed by optimization. *Int. J. Artif. Intell.* **2017**, *15*, 17–29.

20. Butt, C.B.; Hoque, M.A.; Rahman, M.A. Simplified fuzzy-logic-based MTPA speed control of IPMSM drive. *IEEE Trans. Ind. Appl.* **2004**, *40*, 1529–1535. [CrossRef]

21. Hossain, M.S.; Hossain, M.J. Performance analysis of a novel fuzzy logic and MTPA based speed control for IPMSM drive with variable d- and q-axis inductances. In Proceedings of the 12th International Conference on Computers and Information Technology, Dhaka, Bangladesh, 21–23 December 2009; pp. 361–366.

22. Uddin, M.N.; Rebeiro, R.S. Online efficiency optimization of a fuzzy-logic-controller-based IPMSM drive. *IEEE Trans. Ind. Appl.* **2011**, *47*, 1043–1050. [CrossRef]

23. Cao, X.; Fan, L. A Novel Flux-weakening Control Scheme Based on the Fuzzy Logic of PMSM Drive. In Proceedings of the 2009 IEEE International Conference on Mechatronics and Automation, Changchun, China, 9–12 August 2009; pp. 1228–1232.

24. Uddin, M.N.; Chy, M.M.I. A Novel Fuzzy-Logic-Controller-Based Torque and Flux Controls of IPM Synchronous Motor. *IEEE Trans. Ind. Appl.* **2010**, *46*, 1220–1229. [CrossRef]

25. Gethertech Inc. Available online: https://www.gathertech.net/hil-product (accessed on 15 April 2018).

26. Besançon-Voda, A. Iterative auto-calibration of digital controllers: Methodology and applications. *Control Eng. Pract.* **1998**, *6*, 345–358. [CrossRef]

27. Precup, R.-E.; Preitl, S.; Faur, G. PI predictive fuzzy controllers for electrical drive speed control: Methods and software for stable development. *Comput. Ind.* **2003**, *52*, 253–270. [CrossRef]

28. Ginter, V.J.; Pieper, J.K. Robust gain scheduled control of a hydrokinetic turbine. *IEEE Trans. Control Syst. Technol.* **2011**, *19*, 805–817. [CrossRef]

29. Jin, Q.B.; Liu, Q. IMC-PID design based on model matching approach and closed-loop shaping. *Isa Trans.* **2014**, *53*, 462–473. [CrossRef] [PubMed]

30. Ellis, G. *Control System Design Guide*, 2nd ed.; Academic Press: San Diego, CA, USA, 2000; pp. 105–106, ISBN 2-12-237465-7.

electronics

MDPI

Article

Hardware Implementation for an Improved Full-Pixel Search Algorithm Based on Normalized Cross Correlation Method

Guohe Zhang [1], Zejie Kuang [1], Sufen Wei [1,3], Kai Huang [1], Feng Liang [1] and Cheng-Fu Yang [2,*]

[1] School of Electronic and Information Engineering, Xi'an Jiaotong University, Xi'an 710049, China; zhangguohe@xjtu.edu.cn (G.Z.); kuangzejie@stu.xjtu.edu.cn (Z.K.); weisufen@jmu.edu.cn (S.W.); huangkailnn@126.com (K.H.); fengliang@xjtu.edu.cn (F.L.)

[2] Department of Chemical and Materials Engineering, National University of Kaohsiung, No. 700, Kaohsiung University Rd., Nan-Tzu District, Kaohsiung 811, Taiwan

[3] School of Information and Engineering, Jimei University, Fujian 361021, China

* Correspondence: cfyang@nuk.edu.tw; Tel.: +886-91-7270840

Received: 7 November 2018; Accepted: 8 December 2018; Published: 12 December 2018

Abstract: Digital speckle correlation method is widely used in the areas of three-dimensional deformation and morphology measurement. It has the advantages of non-contact, high precision, and strong stability. However, it is very complex to be carried out with low speed software implementation. Here, an improved full pixel search algorithm based on the normalized cross correlation (NCC) method considering hardware implementation is proposed. According to the field programmable gate array (FPGA) simulation results, the speed of hardware design proposed in this paper is 2000 faster than that of software in single point matching, and 600 times faster than software in multi-point matching. The speed of the presented algorithm shows an increasing trend with the increase of the template size when performing multipoint matching.

Keywords: full pixel search algorithm; digital speckle correlation measurement method; hardware implementation

---

1. Introduction

With the rapid development of advanced manufacturing technology, the research of three-dimensional measurement technology is widely used in the fields of material testing, strength testing, and quality control. Digital speckle correlation measurement is an important method in the field of modern optical measurement [1]. Compared with other optical measurement methods, the most significant advantage lies in its simple experimental equipment and convenient measurement. It has the characteristics of full field non-contact, no damage, high accuracy, and high automation by using binocular image acquisition equipment [2–4]. Because of these characteristics, the digital speckle correlation method breaks the limitations of traditional measurement methods and creates a wide range of applications, such as the measurement of deformation of composite materials [5], the detection of cancer cells by measuring the skin strain [6], and measurement in a complex and severe environment [7–10], et al.

The basic idea of digital speckle correlation method is to divide the search process into two parts, the full-pixel search and the sub-pixel search [11]. The full-pixel search is performed in integer pixels, and rapidly locates in a large area. The sub-pixel search is done on the basis of full-pixel to further improves the accuracy. Most applications require a precision in the order of 1% of a pixel, and only the sub-pixel search can achieve this accuracy requirement. Obviously, the full-pixel search is very essential to the feasibility, accuracy, and high speed of subsequent sub-pixel search. The low speed during the matching search is the main problem needed to be solved especially in the case of high computational complexity and large amount of data processing [12]. Normalized cross correlation (NCC) method based on gray features is used as the similarity measure function in the image matching search [13]. It has the characteristics of high accuracy, good performance such as anti-noise and adaptability to image distortion. However, problems such as complex correlation computation, low computation speed, and long computation time still exist with software implementation. The limitations are unsuitable for some real-time system applications. Nowadays, the hardware implementation of image algorithms has been widely applied in the field of image processing with the advantages of high degree of parallelism, integration and resulting high speed, low power consumption, and low cost [14,15]. The implementation of image algorithms through hardware can overcome the computational bottleneck and improve the processing speed of the system.

Here, an improved fast full-pixel search algorithm considering hardware implementation is proposed to solve the existing problems of low speed and high complexity in the NCC algorithm. A compromise strategy of rough matching and exact matching is adopted. When the system is implemented in FPGA (Field Programmable Gate Array), the corresponding number of multipliers is used to calculate the correlation coefficient of the local matching template and the time-sharing method is adopted to speed up the full template calculation.

2. Improvement and Hardware Design of Full-Pixel Search Method

2.1. Digital Speckle Correlation Calculation and Correlation Coefficient

The digital speckle correlation matching is to obtain the displacement and the deformation of an object according to the cross correlation of the speckle images before and after the object is deformed. The digital speckle correlation calculation process is shown in Figure 1.

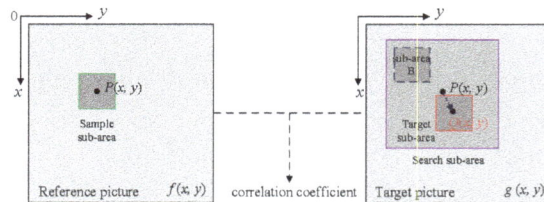

Figure 1. Calculation procedure of Digital Speckle Correlation Method.

The gray scale distribution function of a reference image before deformation is set to $f(x,y)$. The distribution function of a target image after deformation is set to $g(x,y)$. In the reference picture, a rectangular area of size $N \times N$ centered on a point $P(x,y)$ to be measured is selected as a sample sub-area which usually become a matching template. At the same time, in the target picture, a rectangular area of size $M \times M$ (M is greater than N) centered on $P(x,y)$ is selected as a search sub-area which usually become a matching window. Then correlation operations are performed in the search sub-area with the sub-area B, which is of the same size as the sample sub-area, to find the point $Q(x,y)$ corresponding to the extreme value (dependent on the correlation function) of the correlation coefficient of the sample sub-area selected in the reference picture. Thereby, the displacement component of the reference point $P(x,y)$ in the x and y directions can be determined.

The digital speckle correlation method generally uses a gray-based statistical correlation algorithm. Its principle is to determine whether the two sub-regions are related according to the statistical characteristics of the gray distribution of each sub-region before and after the deformation. Usually, the judgment is based on various correlation coefficient functions. This paper adopts normalized cross correlation coefficient [16,17].

2.2. Improved Fast Integer Pixel Search Method

The flow chart of the overall implementation of the algorithm is shown in Figure 2. Firstly, the reference image and the target image are imported. The reference sub-area is selected as the full template of the match in the reference image, and the initial search area is selected in the target image. Then, the local template slides from left to right and from top to bottom in the search area. Correlation function is used to calculate the correlation coefficient value in each sliding process and to compare the calculated correlation coefficient value with the threshold. If the correlation coefficient value is greater than the threshold value, it means that the search window has strong correlation with the reference sub-area and can be used as a candidate matching window. If the correlation coefficient value is less than the threshold, the correlation coefficient value of the next window will be continued to calculate. At the same time, the histogram statistics strategy is used to count the correlation coefficient values in the whole local template matching process. Then a full template is used to calculate the correlation coefficient values of all candidate matching windows to determine the best matching point, and the search task for a pair of matching points is completed. Finally, the correlation coefficient values of the histogram statistics are processed to calculate the optimal threshold of the next pair of matching points. The position and size of the search area of the next matching point are adjusted adaptively by the displacement component of the matching point.

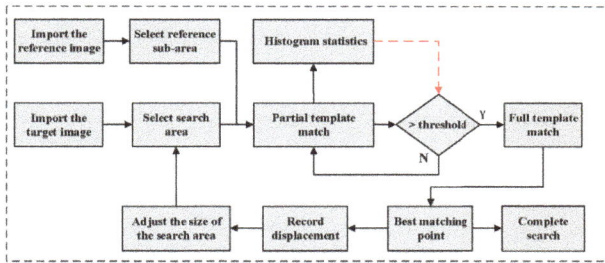

Figure 2. Overall algorithm flow chart.

The improved algorithm mainly includes the following three key steps:

(1)   Selection method of local matching templates

The two-layer matching algorithm is adopted in this paper. Compared with the three-layer matching algorithm [18], the multiple matching tedious steps are reduced. As an important step to improve the calculation efficiency of the algorithm, local template matching not only requires as few pixels as possible, but also needs to reflect the texture information of the full template as much as possible. Therefore, regional blocks are selected at various locations in the full template to form a local matching template. As shown in Figure 3, five regions R0, R1, R2, R3, and R4 were selected to form a local template. R0 is distributed in the middle of the full template which is made up of rectangular areas of $M \times M$ size, and the other four regions are distributed on the four corners of the full template. These five regions are distributed in different positions of the template and can effectively represent the texture information of the entire template. If the size of the local template is larger, the accuracy of the matching is higher, but correspondingly, the consumption of resources is greater. When assuming the entire template as a rectangular area of size $H \times H$, a rectangular region of $h \approx H/4$ can be selected as a local matching template considering the accuracy and resource consumption. If a 31×31 full template is used, the edge length of the local template can be selected as 7 pixels. The purpose of the local template method is to exclude some non-matching points and thus reduce the computational complexity. The best match point is obtained after the full template calculation.

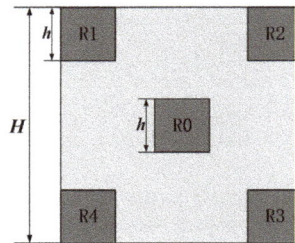

Figure 3. Template for local matching.

(2) Adaptive selection method of thresholds for histogram statistics

We use the correlation coefficient of the local template as the threshold. The closer the correlation coefficient is to 1, the stronger the correlation between the two sub-areas (reference sub-area and search sub-area) is; so is the threshold. The choice of threshold is extremely important to the whole algorithm. The correlation coefficient calculated by the local template needs to be compared with the threshold to exclude some poorly correlated points. If the selected threshold is too small, a large number of matching points will be retained, and the total number of full template calculations will be increased. If the selected threshold is too large, the best matching point might be eliminated which will be resulting in a greater likelihood of image mismatch.

In order to improve the reliability and flexibility of the algorithm, an adaptive threshold selection method is presented. Image correlation matching algorithm usually needs to perform ergodic correlation operations on all pixels of the speckle image before and after the entire displacement to determine the displacement vector of the whole field pixel (as shown in Figure 4). There is an overlapping area between the reference sub-area and the search area of two adjacent matching points. The texture information, the noise, and the exposure effects of them are similar. Therefore, a certain similarity can be seen in the distribution of their correlation coefficients. Based on this feature, the distribution of correlation coefficients of the current matching points under the local matching template can be collected to obtain the number of correlation coefficient values in each interval. The appropriate correlation coefficient value can be calculated as the threshold of the next matching points. The overlapping search areas of two adjacent points A and B to be matched are shown in Figure 5. We use the histogram to make statistics on the correlation coefficient of matching point A under the local template. The correlation coefficient interval 0~1 is subdivided into 100 intervals as the abscissa of the histogram, and the length of each interval is 0.01. The ordinate of the histogram represents the number of matching points in a certain interval. When the local matching template is swiped once in the search area, a correlation coefficient is calculated. The interval is determined by where the correlation coefficient is located. The histogram of the corresponding interval plus 1 is generated to complete the statistics of the correlation coefficient. When the matching point search is completed, the number of occurrences of the correlation coefficient in each interval can be obtained. After completing the statistics, the number of occurrences is sequentially accumulated from the larger interval to the smaller interval of the histogram. So, accumulation stops when the accumulated value reaches 8% of the total amount, and the current abscissa is used as the search threshold of matching point B. Because the full template only needs to participate in the calculation of about 8% window in the search area, a large amount of non-matching points can be excluded by this method. It greatly reduces the amount of calculations, saves search matching time, and improves the matching efficiency of the algorithm.

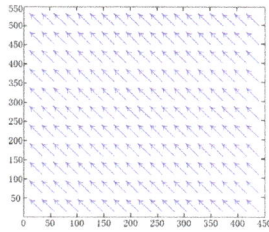

Figure 4. Full field whole pixel displacement vector.

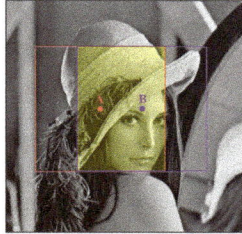

Figure 5. Overlapping regions of adjacent matching points.

(3)   Adaptive adjustment of the location of the search areas

When industrial measurements are made using digital speckle, the displacement or deformation of each point on the surface of a object can be considered to vary continuously. In general, sudden mutations are less likely to occur, so it can be considered that the displacement values of adjacent two matching points in the speckle pattern before and after the displacement are not significantly different. In the traditional full-pixel search algorithm, a rectangular area which is centered on the coordinates of the center point of the reference sub-region and larger than the reference sub-area is framed in the target image as the search area. Therefore, in the case of uncertain object displacement, it is necessary to select a larger search area to prevent the best matching point from image mismatch. However, if the search area is too large, the number of searches increases and the search speed decreases.

In order to solve the above problems, a method for adaptively adjusting the search area is presented, as shown in Figure 6. Assuming the point P in the first picture is the initial point to be matched of the reference image, a larger rectangular area is selected as the search area P centered on the coordinates of the point P in the target image, as shown in the second picture. When the search is completed, the best matching point P^* is obtained, and the displacement (u, v) of the current matching point is recorded. When searching for the adjacent matching point Q, Q^* corresponding to point Q after the displacement (u, v) is found in the target image. Then, taking point Q^* as the center, a rectangular area smaller than the search area P is selected as the search area of the Q point, as shown in the third picture. The size of the initial search area and the adjusted search area are configurable. In the case of sudden mutations, the size of the adjusted search area can be appropriately increased. Search for the best match point of the Q point in this search area to complete the search. The displacement of the point Q is recorded, and the next search area is adjusted with it. The method improves the flexibility of the search area selection, further reduces the search time and improves the efficiency of the algorithm matching search by adjusting the position and size of the search area adaptively.

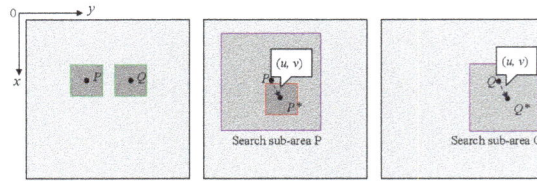

Figure 6. Adaptive adjustment of search areas.

3. Hardware Circuit Design

The full-pixel search matching is a crucial step as well as the most complex step in the digital speckle correlation method. Considering the improvement of resource utilization and the matching speed, a hardware implementation circuit of integer pixel fast matching algorithm is designed.

Through the improvement of the matching search algorithm, the fast matching algorithm proposed in this paper can be better suited to the hardware implementation. The structure of the hardware implementation is shown in Figure 7. The buffer unit reads in the data of the reference image and the target image serially and outputs them parallelly. The algorithm is implemented by using the corresponding number of multipliers to calculate the correlation coefficient of the local matching template firstly. If the full template calculation is required after the threshold comparison, the multiplier array is multiplexed by time-sharing method. It ensures that the full template calculation for this window is completed in three clock cycles. The cost of hardware implementation is greatly reduced.

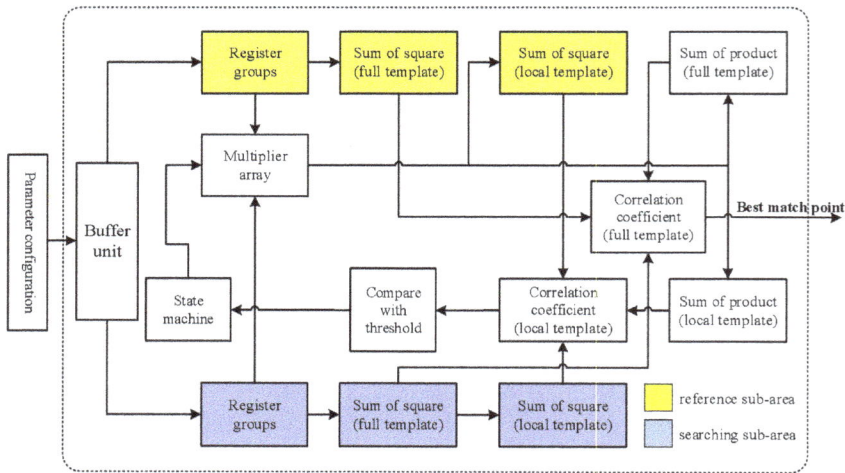

Figure 7. Structure of hardware implementation.

The hardware implementation adopts the matching template of size 31×31. The pixel size of the reference image and the search image is 512×512, the search area size of the initial point is 256×256, and the size of the search area after the adaptive adjustment is 151×151.

3.1. Buffer Unit

(1) Data preprocessing module

In this paper, two Synchronous Dynamic Random Access Memories(SDRAMs) are used as storage units for the reference image and the target image. The structure of the SDRAM control unit is shown in Figure 8.

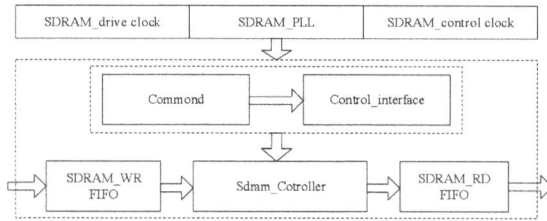

Figure 8. Structure of the SDRAM control unit.

The SDRAM requires a drive clock and control clock. A fixed phase difference between the drive clock and the control clock is required to ensure stable read/write data. Therefore, a PLL is added to the design. The external clock separates two clocks with the same frequency and different phases through the PLL. The sequential control of the SDRAM is composed of the Command module and the Control_interface module. The Command module is primarily responsible for the control of precharge, refresh and burst read/write. Control_interface module controls the operation of SDRAM directly. First, the SDRAM is initialized. Then the SDRAM performs normal read/write operations. All of the rows and columns of Banks must be pre-refreshed every 64 ms to ensure that data is not lost. The read/write dual ports are the SDRAM_WR_FIFO input port and the SDRAM_RD_FIFO output port. Since the read/write operations of the SDRAM share a set of I/O ports, the SDRAM can only perform the read operation or the write operation at the same time.The asynchronous FIFOs are used to achieve cross-clock domain interaction with the data. The control operation of the FIFOs is as follows:

When a write request occurs, the data is sent to the WR_FIFO for caching;

When a read request occurs, the data is sent to the RD_FIFO for caching;

When the data in the WR_FIFO exceeds a certain depth, the SDRAM write operation is requested, and the burst write operation is started when the response is received;

When the data in the RD_FIFO is less than a certain depth, the SDRAM read operation is requested, and the burst read operation is started when the response is received.

(2)   Parallel data output module

The parallel data output module reads the data of the reference sub-area and the search sub-area and performs serial-to-parallel processing on the data of the search sub-area. Figure 9 shows the timing of the reference sub-area data output. First, send a command to SDRAM to read the data of the reference sub-area (31 rows, 31 columns) in the burst mode. Ref_Data_En is the enable signal of the row. Every enable single contains 31 valid data. Then, read the data of the next row after a certain time interval. The amount of data in the reference sub-area is significantly less than that in the search area. The read time of the reference sub-area data is shorter than the read time of the search area data. Therefore, the data output to the reference sub-area can be taken in a serial way.

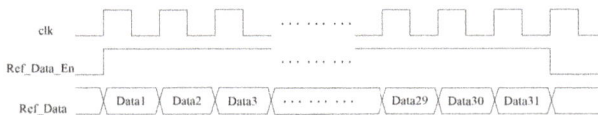

Figure 9. The timing of the reference sub-area data output.

Figure 10 shows the timing of the search sub-area data output. It is similar to the reference sub-area.

Figure 10. The timing of the search sub-area data output.

Then, the data is sent to the buffer unit for serial to parallel processing. As shown in Figure 11. It contains 31 sets of dual port Rams (Dpram) with a data depth of 256 and a data bit width of 8 bits. The Dprams are connected in a serial way. When the high level of the RD_FIFO_Enable signal is detected, the read address of the buffer unit is incremented by 1 and is used as the write address of the next clock cycle. The Search_Data is delayed by one clock and then connected to the Din of the Dpram. When the read address is increased to Size_SR-1, it indicates that all the row data in the Dpram has been read. At this time, clear the read address immediately and continue reading the data.

Figure 11. Dual port Rams (Dprams) of the search sub-area.

The data of a row will be read in one clock cycle. When 31 rows of data are read, the output of the port Dout 1~31 of the buffer unit is the parallel input of the first 31 rows of the search area. The 31 sets of parallel data are valid data for correlation matching operations.

(3) Matching window

The data of the reference sub-area remains unchanged throughout the search process, and its amount of data is small. Once the position of the reference sub-area is determined, the data is immediately read from the SDRAM and stored in the register group. As shown in Figure 12, it is a register group consisting of 31 × 31 registers. Every register can store the 8-bit data of grayscale image.

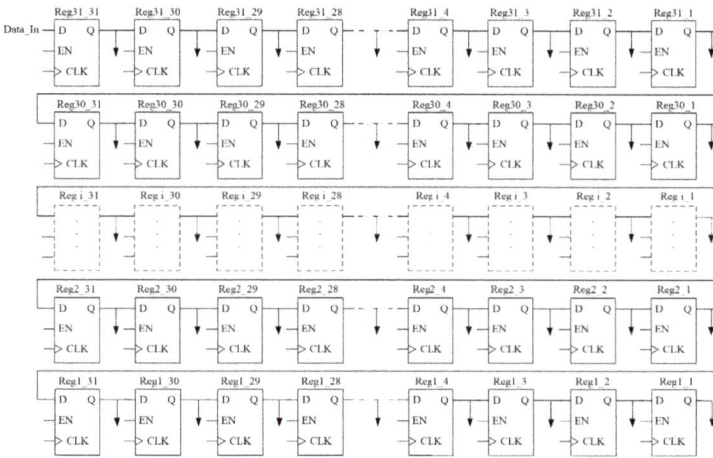

Figure 12. Serial shift register groups of the reference sub-area.

For the sliding window of the search sub-area, the module consists of 31 parallel shift register groups. The hardware implementation is shown in Figure 13. The input of the register group is 31-Channel 8-bit parallel output of the Dpram. The 31 × 31 parallel shift register group is followed by a First Input First Output buffer(FIFO) and another set of 31 × 31 parallel shift registers. The result of calculating the correlation coefficient of the local matching template requires a delay of several tens of clock cycles. During these tens of clock cycles, the data of the previous 31 × 31 parallel shift register group will be continuously updated. When the window is judged to require full template matching, the data of the current window is updated early and the full template calculation cannot be performed. Therefore, the FIFO is used to cache the data of the parallel shift register group. Register groups in the solid wireframe are used to calculate the correlation coefficients of local templates. Register groups in the dotted wireframe are used to calculate correlation coefficients of full templates.

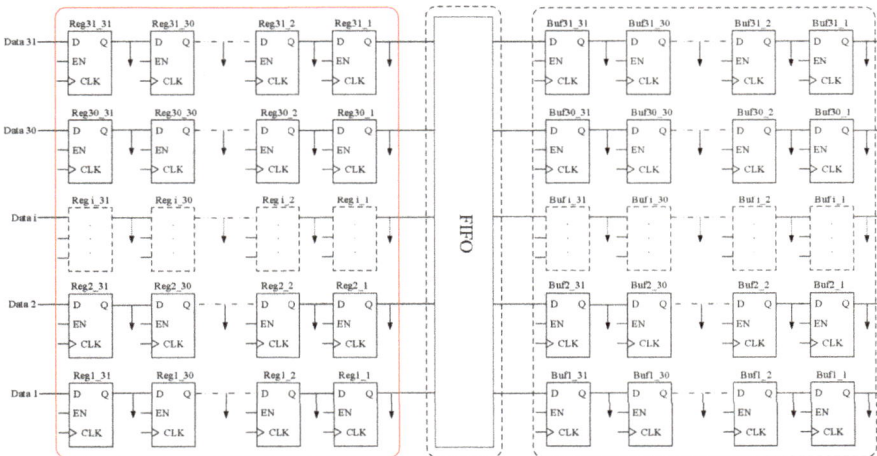

Figure 13. Serial shift register groups of the search sub-area.

3.2. Sum of Squares of All Pixels in a Reference Sub-Area

The data of the reference sub-area of a matching point is fixed as a matching template throughout the search process. Therefore, each matching point only needs to calculated once.

The circuit structure shown in Figure 14 is the sum of squares calculation unit of the reference sub-area. It accumulates the data of the search sub-region one by one while the reference sub-region data enters the serial shift register group.

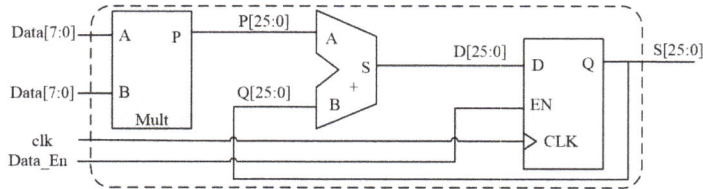

Figure 14. Sum of squares calculation unit of the reference sub-area.

3.3. Sum of Squares of All Pixels in a Search Sub-area

The corresponding local matching template is divided as shown in Figure 3. The size of the full template is 31×31, and R0, R1, R2, R3, and R4 are all rectangular areas of size 7×7. Different from the reference sub-area, the data in the search sub-area register group is dynamically changed. Therefore, the parallel pipeline structure is used to calculate the sum of squares of the pixels in the search sub-area. The five regions in the local template are divided into three parts for parallel calculation, which are the sum of squares of the regions R1R2, R0, and R3R4.

Figure 15 shows the circuit for calculating the sum of squares of R1R2. The circuit implements the sum of squares calculation of the first 7 lines data in the template and the sum of squares of the region R1R2 in a single clock cycle. First, the 7-way parallel data is calculated by a 8-bit multiplier for the square of each data. Then the summation calculation is done by the pipeline adder. The result is serially stored in 31 20-bit shift registers. Finally, the sum of squares of the region R1R2 (Sum_R1R2) is calculated by the pipeline adders 1, 2 and the adder A, and the sum of squares of all 7 rows (Sum1) in the window can be calculated by the pipeline adder 3 and the adder B.

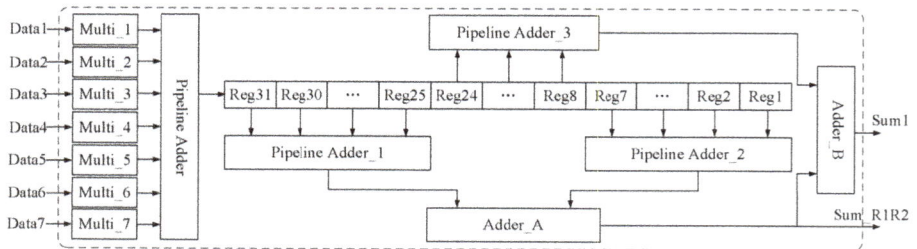

Figure 15. Calculating the sum of squares of R1R2.

A similar circuit structure is used to obtain the sum of squares of the region R0, R3R4 and the 7 rows occupied by them, and the sum of squares of the remaining 10 rows. The sum of squares of a local template can be obtained by summing Sum_R1R2, Sum_R0, and Sum_R3R4. The sum of squares of the full template can be obtained by summing Sum1, Sum2, Sum3, and Sum4. Hence, they will be ready for full template matching which is possible later.

3.4. Sum of Cross-Correlation Product

For the sum of cross-correlation product (sum of product of the pixel in a reference sub-area and the corresponding one in a search sub-area), a large amount of multiplier resources are required to ensure that the calculation is to be completed in one clock cycle. Because of the strategy of the local template matching, the resource consumption of Digital Signal Processing(DSP) slices is reduced significantly at the cost of a little speed loss.

As shown in Figure 16, the hardware structure constructs a multiplier array with the same number of multipliers as the local template pixels, and the reference sub-area and search sub-region register groups are treated as inputs of the multiplier array. In this way, the multiplier array can ensure that the cross-correlation product calculation of all pixel points in the local template is completed in one clock cycle. Finally, the sum of cross-correlation product of the local template can be obtained by summing all product by the pipeline adder. For the pipeline adder, cycles consumed to calculate the first data depend on the series of the pipeline. Then, each calculation would be completed in one clock cycle.

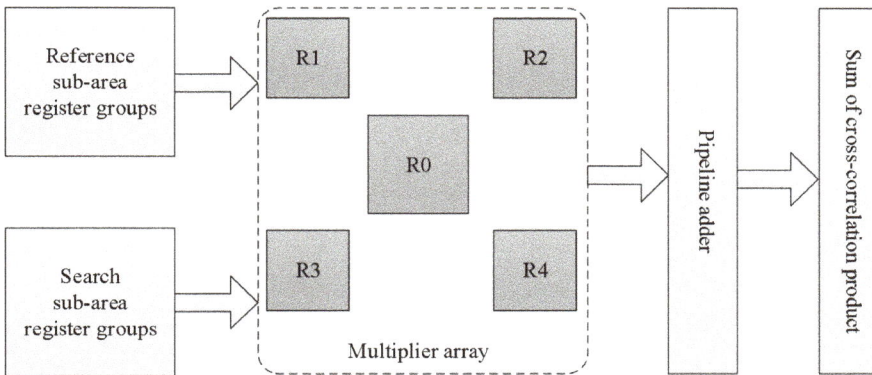

Figure 16. Calculating the sum of cross-correlation product.

3.5. Sum of Product of the Full Template

For the calculation of the correlation coefficient of the full template, we have obtained the sum of squares of the reference sub-areas and search sub-areas. What needs to be calculated is the sum of the cross-correlation products of the full template. However, the calculation of all cross-correlation product cannot be completed in one clock cycle with existing multiplier arrays. Therefore, this paper uses the multi-clock cycle strategy to calculate the cross-correlation product of the full template. As shown in Figure 17, the remaining pixels of the full template are divided into three parts. When the size of the template changes, the system will still calculate the sum of squares of the reference sub-area and calculate the sum of cross-correlation product correctly. What we need to do is adjust the regional division for time-sharing calculation to ensure that the multiplier arrays can complete full template calculations in three clock cycles.

When the full template correlation coefficient needs to be calculated, the system will make the full template calculation flag to be equal to 1. This signal will cause the cache unit to stop outputting data. The data of the parallel shift register group is held for three cycles, so that the multiplier array is idle to process the product of the full template data. The calculation would be completed in three cycles. Figure 18 shows the state machine that controls the multiplier array. The operation of each state is shown in Table 1.

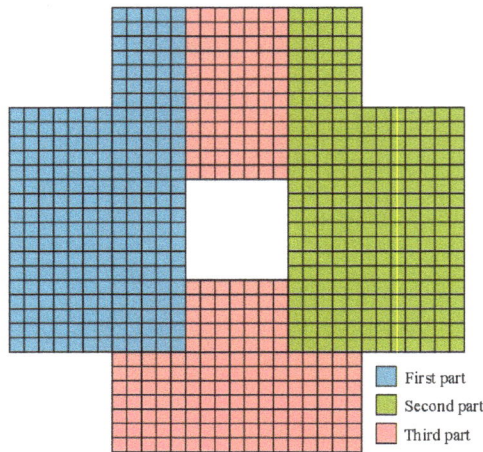

Figure 17. Regional division for time-sharing calculation.

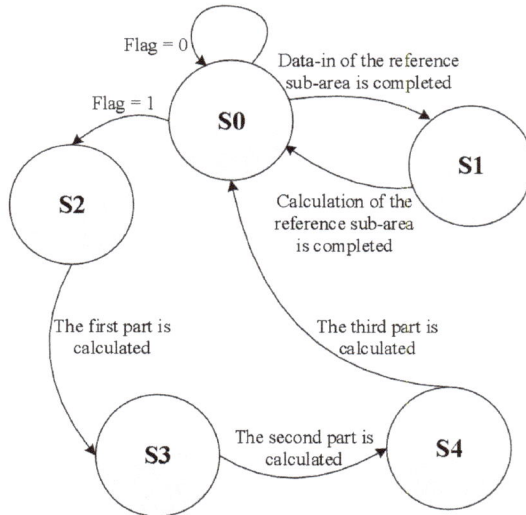

Figure 18. State machine that controls time-sharing calculations.

Table 1. Resource consumption in field programmable gate array (FPGA).

State	Operation
S0	Calculate the product of the search sub-region and the reference sub-region of the local template.
S1	Calculate the square of the reference sub-region of the local template.
S2	Calculate the product of the first part of the full template.
S3	Calculate the product of the second part of the full template.
S4	Calculate the product of the third part of the full template.

3.6. Selection of Adaptive Threshold

Under ideal conditions, the best matching point should be the one with the correlation coefficient closest to 1. There may be some deviations in consideration of the effects of noise and exposure. But the best match point is still one of the points where the threshold is closest to 1. We count all points with the correlation coefficient from 1 to 0 by the histogram and retain those points close to 1. As shown in Figure 19, the work of Dpram is to count the number of correlation coefficients in different intervals. 1~100 address values of Dpram represent the correlation coefficient value of 0.01~1 . The calculated correlation coefficient multiplied by 100 is used as the read address of Dpram. The value in this address of Dpram is incremented by 1. After the local template completes the calculation of the entire search area, the values in Dpram are taken one by one to be summed from the largest address to the minimum address. Accumulation stops when the accumulated value reaches 8% of the total amount, and the correlation coefficient corresponding to the current address is used as the threshold. This percentage value is configurable. It depends on the specific application scenario. Its determinants are mainly noise and exposure. If the noise or the exposure is high, it can be increased to include more points. Inevitably, increasing this value will result in an increase in the amount of calculation. In most cases, such as the conditions of the experiments in this paper, 8% is sufficient.

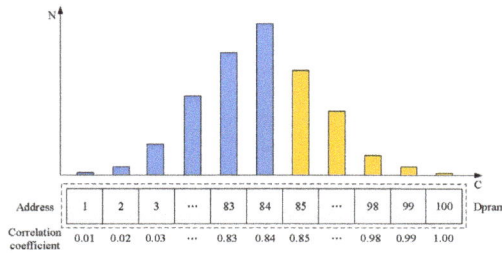

Figure 19. Dpram for histogram statistics.

4. Simulation and Verification of Hardware Design

4.1. Simulation Results and Analysis

The size of the reference sub-area (calculation window) is an important factor affecting the accuracy of the entire pixel search. If the calculation window is too small, the amount of information that contains image features is small. Although the calculation speed will be increased, the correlation coefficient between adjacent points will have obvious mutations. Correspondingly, if the size of the sub-region is too large, the amount of information that contains image features is large, which can compensate for the effects of various introduced noises, but the calculation time also increases. Therefore, choosing a suitable size of the calculation window is a prerequisite to ensure the correctness of other series of experiments. In this section, the influence of the calculation window on the correlation coefficient is analyzed through experiments. The experiment collected two speckle patterns before and after the deformation by charge coupled device(CCD) camera. The speckle pattern pixel size is 1080×800. The horizontal displacement and vertical displacement of the full-pixel have been determined in advance. A search window of a different size is selected in the reference image for search in the target image. The correlation coefficients are calculated to obtain a correlation coefficient grid maps for several different calculation windows as shown in Figure 20.

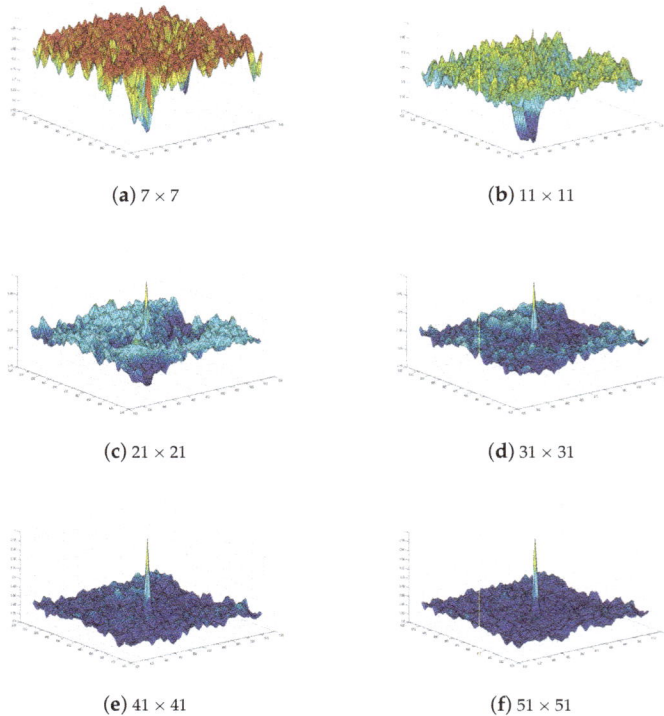

(**a**) 7×7

(**b**) 11×11

(**c**) 21×21

(**d**) 31×31

(**e**) 41×41

(**f**) 51×51

Figure 20. Correlation coefficient grid diagram of different sizes.

From the above grids, we can conclude that when the size of the calculation window is less than 31×31, the fluctuation of the correlation coefficient is relatively large. The correlation coefficient has a small difference from other peaks. The main peak is not obvious, and it is easily disturbed by noise and other factors. When the size of the calculation window is above 31×31, the fluctuation of the correlation coefficient is relatively flat, and as the calculation window increases, the variation range of the correlation coefficient also tends to be gentle and the accuracy of the calculation also increases. As the calculation window continues to increase, the correlation calculation amount also increases sharply, however, the calculation accuracy does not increase significantly as the calculation window increases. Therefore, it is appropriate to select a search window size of 31×31 or 41×41. The time for search is relatively short, while ensuring accuracy. Two sets of experiments are performed on the proposed fast algorithm. In the first set of experiments, the template size is 31×31 and the search area size is 256×256. In the second set of experiments, the template size is 61×61 and the search area size is 256×256. Each group of experiments conducts a search for different numbers of matching points. The time consumed for search with different number of matching points is counted and compared with other algorithms. The results are shown in Table 2.

Table 2. Comparison of time consumption for three matching algorithms.

Algorithm	Search Area Size	Template Size	Search Time (s)		
			1 Point	64 Points	100 Points
NCC [19]	256×256	31×31	4.04	221.56	322.27
		61×61	15.96	1087.53	1523.28
NCC-BF [20]	256×256	31×31	2.63	159.26	228.56
		61×61	12.65	801.34	1286.25
This paper	256×256	31×31	1.36	14.87	22.76
		61×61	5.01	69.15	101.41

As shown in Table 2, we can conclude that the speed of the algorithm proposed in this paper is slightly better when performing single-point matching, which is about three times that of the NCC algorithm [19] and twice that of the NCC-BF algorithm [20]. When the template size is 31×31, its matching speed is about 15 times that of the NCC algorithm and 11 times that of the NCC-BF algorithm. When the template size is 61×61, the matching speed is about 16 times that of the NCC algorithm and 12 times that of the NCC-BF algorithm. Therefore, the methods of adaptive threshold and adaptively adjusting the search areas adopted by this algorithm are suitable for multi-point matching search. The larger the template is, the greater the advantage is.

4.2. FPGA Verification Results

This experiment uses the Stratix IV series FPGA of company Altera. Using Quartus II to synthesize the sub-pixel search module, the circuit resource obtained through the auto place & route tool is shown in Table 3.

Table 3. Resource consumption in FPGA.

Resource	Consumption
Combinational Adaptive Look-Up Tables(ALUTs)	8.55%
Dedicated logic registers	5.96%
Total pins	6.08%
Global Clock Buffer(BUFG)	18.75%
Total block memory bits	0.35%
DSP block 18-bit memory bits	28.42%

As shown in the Table 3, the hardware consumption is relatively small while ensuring higher speed. In FPGA verification, the coordinates of the matching points of the full-pixel search and sub-pixel search [21] are input to the serial port debugger through the serial cables and saved as text information. Comparing this result with the results of MATLAB, the two are completely consistent, indicating that the circuit design of the digital speckle correlation method is correct on the FPGA.

The operating time obtained by the logic analyzer is shown in Table 4. The results show that the speed of hardware design proposed in this paper is about 2000 times faster than software when performing single-point matching. The speed of hardware is 600 times faster than the software when performing multi-point matching. It only takes 520 microseconds when the hardware performs a single-point match. It's only a few tens of milliseconds when doing a 100-point match. Therefore, the hardware design of this article achieves the effect of real-time processing.

Table 4. Comparison of simulation time between hardware and software.

Platform	Experiment Platform	Search Area	Template Size	Matching Time		
				1 Point	64 Points	100 Points
Software	Intel-i5 2400 3.1GHz	256×256	31×31	1.36 s	14.87 s	22.76 s
FPGA	Altera Stratix IV	256×256	31×31	0.52 ms	23.09 ms	37.62 ms

5. Conclusions

In this paper, a fast full-pixel search method suitable for digital speckle correlation method is proposed based on NCC correlation matching algorithm. The method first uses the local matching template to eliminate the number of non-matching points to avoid the calculation of unrelated points, thereby improving the calculation speed and efficiency of the algorithm. Then, the histogram statistics are used to select the threshold adaptively. The effects of image noise and local exposure on the algorithm are reduced, which improves the flexibility and robustness of the algorithm. Later, the adaptive adjustment of the search area is used to adjust the position and size of the search area. The calculation speed and computational efficiency are greatly improved without loss of accuracy. Finally, a hardware circuit implementation is realized for the fast integer pixel search algorithm. The circuit can be applied to a real-time image processing system with small hardware cost while ensuring a small loss of search speed.

Author Contributions: Conceptualization, G.Z.; Data curation, Z.K. and K.H.; Formal analysis, G.Z.; Funding acquisition, F.L.; Investigation, S.W.; Methodology, S.W.; Project administration, C.-F.Y.; Resources, C.-F.Y.; Validation, Z.K.; Writing—original draft, K.H.; Writing—review and editing, Z.K.

Funding: This research was funded by the National Natural Science Foundation of China under Grant 61474093.

Conflicts of Interest: The authors declare no conflict of interest.

References

1. Dong, H.; Zhou, Y.; Guo, J.; Zhang, W.Y.; Wang, W.S. Digital Speckle Pattern Interferometry for Deformation Measurement. *Acta Photonica Sin.* **2010**, *39*, 19–22. [CrossRef]
2. Chen, Z.X.; Liang, J.; Guo, C. Application of digital speckle correlation method to deformation measurement. *Opt. Precis. Eng.* **2011**, *19*, 1480–1485. [CrossRef]
3. Yu, G.; Wang, S.G.; Yu, J.H. Technology of digital speckle pattern interferometry and its applications. *Laser Technol.* **2002**, *26*, 237–240. [CrossRef]
4. Xiao, Q.Y.; Li, J.; Wu, S.J.; Yang, L.X.; Dong, M.L.; Zeng, Z.M. Denoising of DSPI phase map using sine-cosine filtering and signal energy. *Opt. Precis. Eng.* **2018**, *4*, 936–943. [CrossRef]
5. Sun, T.; Liang, J.; Cai, Y.; Wang, Y.Q. Measurement of deformations for copper/aluminum clad under tension with digital speckle correlation. *Opt. Precis. Eng.* **2012**, *20*, 2599–2606. [CrossRef]
6. Krehbiel, J.D.; Lambros, J.; Viator, J.A.; Sottos, N.R. Digital Image Correlation for Improved Detection of Basal Cell Carcinoma. *Exp. Mech.* **2010**, *50*, 813–824. [CrossRef]
7. Lyons, J.S.; Liu, J.; Sutton, M.A. High-temperature deformation measurements using digital-image correlation. *Exp. Mech.* **1996**, *36*, 64–70. [CrossRef]
8. Anwander, M.; Zagar, B.G.; Weiss, B.; Weiss, H. Noncontacting strain measurements at high temperatures by the digital laser speckle technique. *Exp. Mech.* **2000**, *40*, 98–105. [CrossRef]
9. Hu, Y.; Wang, Y.H.; Bao, S.Y.; Hu, H.R.; Yan, P.Z. Optimal imaging of digital image correlation speckle under high temperature. *Chin. Opt.* **2018**, *11*, 728–735. [CrossRef]
10. Wu, S.J.; Yang, J.; Pan, S.Y.; Li, W.X.; Yang, L.X. Dynamic Deformation Measurement of Discontinuous Surfaces Using Digital Speckle Pattern Interferometry and Spatiotemporal Three-dimensional Phase Unwrapping. *Meas. Sci. Technol.* **2018**, *2*, 121–129. [CrossRef]
11. Pan, B.; Qian, K.; Xie, H.; Asundi, A. Two-dimensional digital image correlation for in-plane displacement and strain measurement: A review. *Meas. Sci. Technol.* **2009**, *20*, 152–154. [CrossRef]

12. Wang, Y.H.; Liang, H.; Wang, S. Advance in digital speckle correlation method and its applications. *Chin. J. Opt.* **2013**, *6*, 470–480. [CrossRef]

13. Sun, M.L.; Zong, G.H.; Bi, S.S.; Yu, J.J. Characteristics of Independence on Image Gray Level in Pattern Matching Algorithm and its Application. *Acta Photonica Sin.* **2009**, *38*, 435–440. [CrossRef]

14. Singh, S.; Shekhar, C.; Vohra, A. FPGA-Based Real-Time Motion Detection for Automated Video Surveillance Systems. *Electronics* **2016**, *5*, 10. [CrossRef]

15. Bravo-Muñoz, I.; Lázaro-Galilea, J.L.; Gardel-Vicente, A. FPGA and SoC Devices Applied to New Trends in Image/Video and Signal Processing Fields. *Electronics* **2017**, *6*, 25. [CrossRef]

16. Schreier, H.W.; Braasch, J.R.; Sutton, M.A. Systematic errors in digital image correlation caused by intensity interpolation. *Opt. Eng.* **2000**, *39*, 2915–2921. [CrossRef]

17. Huang, H.T.; Fiedler, H.E.; Wang, J.J. Limitation and improvement of PIV. I—Limitation of conventional techniques due to deformation of particle image patterns. *Exp. Fluids* **1993**, *15*, 168–174. [CrossRef]

18. Zhang, X.G.; Wang, M.J.; Wang, Y.J. Three-layer fast matching algorithm based on adaptive selection of threshold. *Opto-Electron. Eng.* **2005**, *32*, 83–85. [CrossRef]

19. Zitová, B.; Flusser, J. Image registration methods: A survey. *Image Vis. Comput.* **2003**, *21*, 977–1000. [CrossRef]

20. Sun, C. Fast Algorithm for Local Statistics Calculation for N-Dimensional Images. *Real-Time Imaging* **2001**, *7*, 519–527. [CrossRef]

21. Kuang, Z.J.; Zhang, G.H.; Liang F. Fast Hardware Implementation of Sub-pixel Search Algorithm for Digital Speckle Correlation Method. In Proceedings of the 2018 3rd International Conference on Electrical, Automation and Mechanical Engineering, Xi'an, China, 24–25 June 2018; Atlantis Press: Paris, France, 2018; pp. 356–360, ISBN 978-94-6252-538-2.

Article

electronics

MDPI

Design and Realization of a Compact High-Frequency Band-Pass Filter with Low Insertion Loss Based on a Combination of a Circular-Shaped Spiral Inductor, Spiral Capacitor and Interdigital Capacitor

Ki-Hun Lee, Eun-Seong Kim, Jun-Ge Liang and Nam-Young Kim *

Radio Frequency Integrated Centre (RFIC), Kwangwoon University, Kwangwoon-ro,
Nowon-gu, Seoul 01897, Korea; lkh1001say@gmail.com (K.-H.L.); 3037eskim@gmail.com (E.-S.K.);
liangjun1991@hotmail.com (J.-G.L.)
* Correspondence: nykim@kw.ac.kr; Tel.: +82-02-940-5071

Received: 31 July 2018; Accepted: 10 September 2018; Published: 12 September 2018

Abstract: In this study, the proposed bandpass filter (BPF) connects an interdigital and a spiral capacitor in series between the two symmetrical halves of a circular intertwined spiral inductor. For the mass production of devices and to achieve a higher accuracy and a better performance compared with other passive technologies, we used integrated passive device (IPD) technology. IPD has been widely used to realize compact BPFs and achieve the abovementioned. The center frequency of the proposed BPF is 1.96 GHz, and the return loss, insertion loss and transmission zero are 26.77 dB, 0.27 dB and 38.12 dB, respectively. The overall dimensions of BPFs manufactured using IPD technology are $984 \times 800 \ \mu m^2$, which is advantageous for miniaturization and integration.

Keywords: bandpass filter; integrated passive device; intertwined spiral inductor; spiral capacitor; interdigital capacitor

---

1. Introduction

With the rapid development of mobile phones in the last few decades, the demand for bandpass filters (BPFs) with high performance and integration has increased. The main function of a BPF in a transmitter is to limit the bandwidth of the output signal to the band of the BPF. This will prevent the transmitter from interfering with other parts. At the receiver, the BPF prevents unwanted frequency signals from passing through and can receive or decode signals in the selected frequency range. BPFs optimize the signal-to-noise ratio and sensitivity by preventing noise in the receiver. Therefore, the performance of RF transmission and reception in the wireless communication system needs to be increased by removing noise signals and unnecessary frequency harmonic components other than the desired frequency band. When considering mobile phones and their products, compact size is an important requirement [1–3]. Most handheld devices have particularly stringent miniaturization requirements to meet market expectations.

For this reason, mobile phones demand BPFs with high performance and integration, as well as a compact size and low weight. According to this trend, researchers had used low temperature co-fired ceramic (LTCC) technology, and it was reported in [4–6]. Liquid crystal polymer (LCP) substrates were reported in [7,8]. However, LTCC and LCP are not only higher in cost and bigger in size, but also capacitance densities are low and process tolerance immature compared to the integrated passive device (IPD) technology. IPD technology has been widely used to realize compact BPFs and achieve these demands. IPD technology has advantages that include mass production of devices, high accuracy and better performance compared with other passive technologies. With IPD technology, it is possible to integrate individual passive components into a microwave device or system. IPD technology can now be applied to applications that use the entire passive device and is already applied to the front-end module of mobile systems. Many functional blocks such as filters, baluns, diplexers, transformers, directional couplers and power dividers in mobile telecommunication systems can be created using IPD technology. To obtain a compact BPF and an inductor with a high Q-factor, Yook et al. realized a BPF with a suspended spiral inductor on the substrate of an anodized aluminum oxide, but its insertion loss and return loss performances needed to be improved [9]. Chia et al. used two coupled spiral inductors integrated on an Al2O3 substrate, but it suffered from a high insertion loss [10].

In this paper, we present a BPF based on a combination of a circular-shaped spiral inductor, an interdigital capacitor and a spiral capacitor. The overall dimensions of the BPF manufactured using IPD technology are 984×800 μm^2 which is advantageous for the telecom industry where miniaturization and integration are required. The loss due to the parallel capacitor between the signal electrode and ground electrodes can be reduced using a GaAs substrate. For applications to mobile phones, the BPF has been designed with a center frequency of 2.0 GHz, an insertion loss of 0.27 dB and a return loss of 26.76 dB. In addition, the transmission zero of 5.22 GHz in the stopband region of the filter suppresses signals above 38 dB to generate a wide stopband response.

2. Materials and Methods

In order to improve the return loss and insertion loss performances, the proposed BPF is used on a GaAs substrate that can accept high speed microelectronics. High inductance can be achieved by fabricating two spiral inductors with SiNx between them. Thus, inductance can be improved by the mutual induction effect. Based on this technology, spiral and interdigital capacitors are connected in a circular intertwined spiral inductor to reduce return loss and insertion loss and to have a high Q-factor. By utilizing the internal space, it is possible to manufacture chips with a compact size. We designed the layout by using Agilent Advanced Design System (ADS) software (Version 2016. 01, Keysight Technologies, Inc., Santa Rosa, CA., USA) for the proposed BPF and simulated the results.

The proposed BPF connects the interdigital capacitor and spiral capacitor in series between the two symmetrical halves of a circular intertwined spiral inductor, as depicted in the schematic in Figure 1a. The substrate used for fabrication was a GaAs substrate with a diameter of 6 inch and a thickness of 200 μm. The dielectric constant εr is 12.85, and the loss tangent is 0.006. A standard IPD manufacturing process is followed for filter fabrication [11].

Figure 1b shows the equivalent circuit of the proposed BPF. R denotes the resistance loss of the circular intertwined spiral inductor of inductance (L) with the substrate-related parameters COX, CSUB and RSUB. Csc and *Rsc* denote the coupling capacitance and loss resistance of the spiral capacitor [12,13]. *Csi* is the feedthrough capacitance. *Ci* and *Ri* represent the total capacitance and loss resistance of the interdigital capacitor. *Cp* represents the capacitance effect due to direct flux from the signal electrode to the ground electrode of the interdigital capacitor [14,15]. As shown in Figure 1c, the circular intertwined spiral inductor is deposited on the GaAs substrate and a 2000 Å-thick SiNx is placed between the two inductors. The inductance of a planar circular spiral inductor can be expressed as follows [16]:

$$L = \frac{\mu_0 n^2 d_{avg} c_1}{2} \left[ln(c_2/\rho) + c_3\rho + c_4\rho^2 \right] \tag{1}$$

According to the expression based on the current sheet expression, $c1$, $c2$, $c3$ and $c4$ are layout-dependent coefficients, and their values are 1.00, 2.46, 0.00 and 0.20 for the circular inductor, respectively. The average diameter davg is 0.5 (dout + din), and dout and din denote the outer and inner diameters, respectively. N is the number of turns, and ρ is the fill ratio, which is expressed as ρ = (dout − din)/(dout + din).

In Figure 1d, the interdigital capacitors can be expressed as follows [17]:

$$C_i = \left[\varepsilon_0 \left(\frac{1 + \varepsilon_s}{2} \right) \frac{K\left(\sqrt{1 + k^2} \right)}{K(k)} + \varepsilon_0 \frac{t}{a} \right] (N - 1)L \tag{2}$$

where ε_0 is the free space permittivity, ε_s is the permittivity of the GaAs substrate and $k = a/b$ and K (k) represent the first type of elliptic integral. N is the number of windings, and L is the length of the interdigital electrode.

Similarly, the spiral capacitor shown in Figure 1e can be expressed as follows [18]:

$$C_{sc} = \left[\varepsilon_0 \left(\frac{1 + \varepsilon_s}{2} \right) \frac{K\left(\sqrt{1 + k^2} \right)}{K(k)} + \varepsilon_0 \frac{t}{a} \right] L_c \tag{3}$$

where ε_0 and ε_s denote the free space permittivity and permittivity of the GaAs substrate. The other part is Lc, and Lc is the total length of the spiral capacitor line.

The equivalent circuit of an intertwined spiral inductor, a spiral capacitor and an interdigital capacitor is shown in Figure 1b. The resonance frequency of the BPF is given as follows:

$$f_0 = \frac{1}{2\pi \sqrt{LC}} \tag{4}$$

where L and C are the total inductance and capacitance of the proposed BPF, respectively.

$$Q = \frac{f_0}{B_{3-dB}}, \quad B_{3-dB} = f_H - f_L \tag{5}$$

The Q-factor determines the bandwidth of the resonator according to the center frequency. A high value of the Q-factor indicates that the energy loss is less than the energy stored in the resonator.

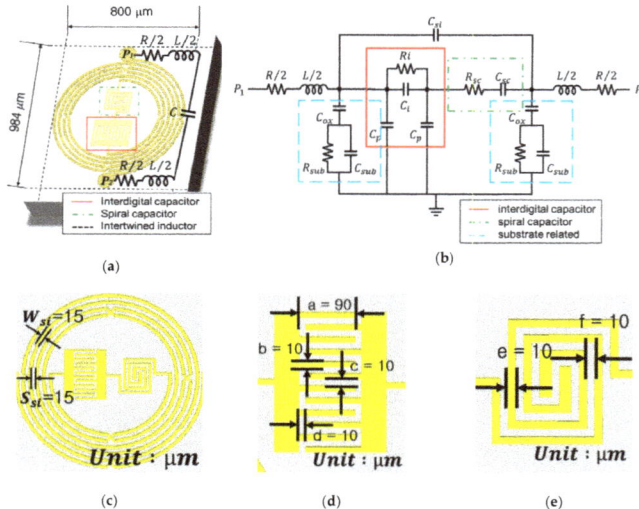

Figure 1. Proposed bandpass filter (BPF) based on integrated passive device (IPD) technology combined circular intertwined spiral inductor, interdigital capacitor and spiral capacitor: (**a**) the layout of the proposed BPF; (**b**) the equivalent circuit of the proposed BPF; (**c**) the layout of the proposed BPF; (**d**) enlarged layout of the interdigital capacitor; (**e**) enlarged layout of the spiral capacitor.

3. Results and Discussion

3.1. Results

The proposed BPF had a center frequency of 2 GHz and was simulated by ADS software to verify it. The center frequency of the BPF obtained from the simulation results was 1.98 GHz, and the return loss, insertion loss and Q-factor were 34.64 dB, 0.06 dB and 1.45, respectively. The optimized dimensions of the circular intertwined spiral inductor, spiral capacitor and interdigital capacitor are as follows.

- The circular intertwined spiral inductor is shown in Figure 1c:
  : $d_{in} = 530$ μm, $d_{out} = 800$ μm, $L_c = 1850$ μm, $N = 5$, $W_{si} = 15$ μm, $S_{si} = 15$ μm
- Interdigital capacitor in Figure 1d:
  : $a = 90$ μm, $b = 10$ μm, $c = 10$ μm, $d = 10$ μm,
- Spiral capacitor in Figure 1e:
  : $e = 10$ μm, $f = 10$ μm, $n = 2$

Figure 2 shows the simulation results of the BPF designed with the above parameters, and the frequency characteristics were confirmed by the S-parameter of the BPF. The frequency characteristics of the resistor, the capacitor and the inductor in Figure 1a are shown.

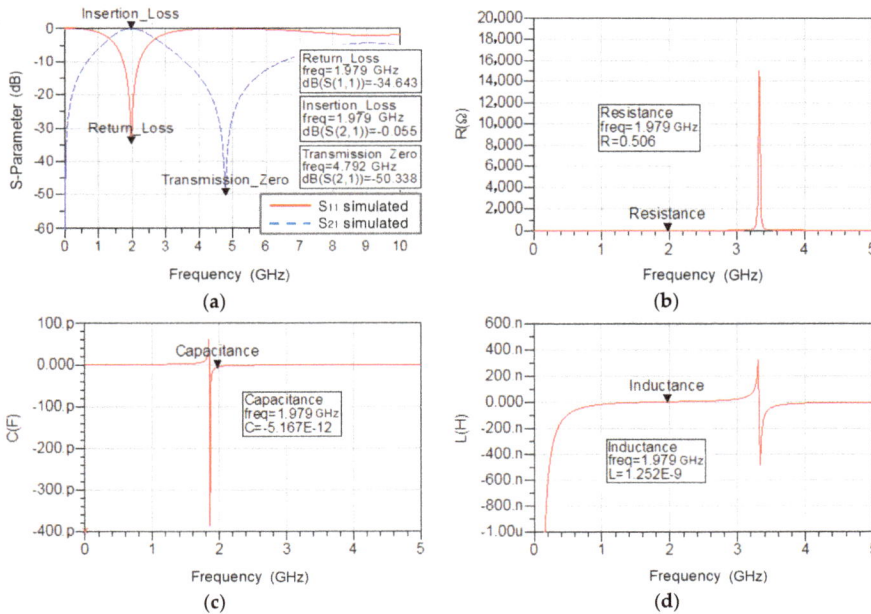

Figure 2. Proposed BPF simulation result to get: (**a**) the S-parameter; (**b**) resistance; (**c**) capacitance; and (**d**) inductance.

The Agilent 8510 vector network analyzer was used to identify the center frequency and characteristics of the fabricated BFP, as shown in Figure 3. The measured center frequency was 1.96 GHz, which was downshifted by approximately 20 MHz from the expected result of 1.98 GHz. The measured return loss, insertion loss and transmission zero were also changed. The measured return loss, insertion loss and transmission zero were 26.77 dB, 0.27 dB and 38.12 dB, respectively. The difference between the simulation results and measured results for the return loss, insertion loss and transmission zero was 7.87 dB, 0.21 dB and 12.22 dB, respectively, and there was a difference between the two results. For transmission zero, the frequency was upshifted to 431 MHz, and the Q-factor changed slightly from 1.45 to 1.41. The reason for the difference between simulation results and measurement results was caused by the dielectric loss of the substrate, the loss in the inductor bending and the accuracy of the physical dimensions. Because of that reason, resistance (from 0.51 Ω–1.74 Ω), capacitance (from 5.17 pF–6.37 pF) and inductance (from 1.25 nH–1.04 nH) also changed. Figure 3b–d displays the scanning electron microscopy (SEM) images of fabricated bandpass filters, interdigital capacitors and spiral capacitors, respectively.

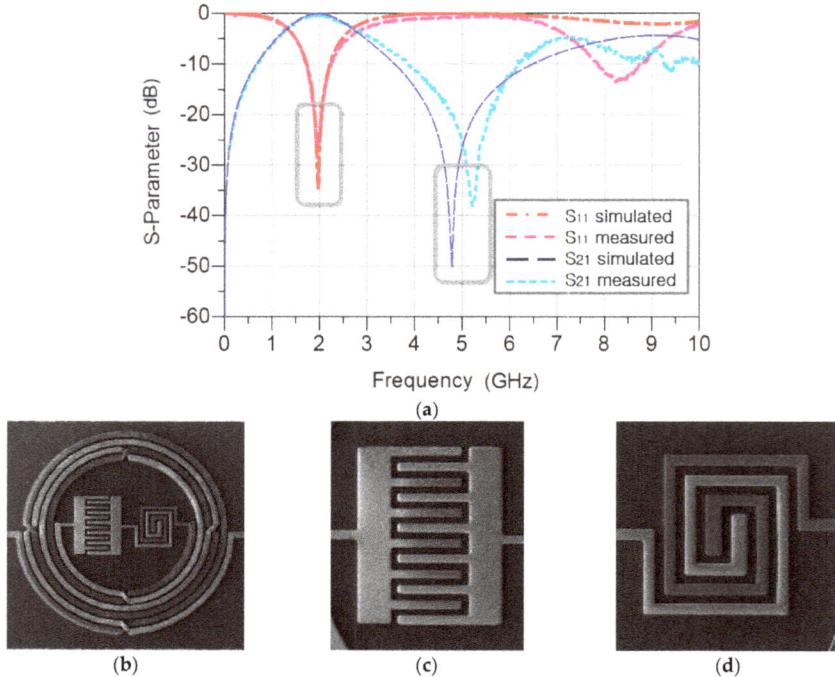

Figure 3. Proposed BPF: (**a**) comparison of simulated and measured results of BPF; (**b**) the scanning electron microscopy (SEM) image of the proposed BPF; (**c**) enlarged SEM image of the interdigital capacitor; (**d**) enlarged SEM image of the spiral capacitor.

3.2. Discussion

The main purpose of this study was to show that the performance and size of filters fabricated through the IPD process were unique and performed better when compared to other works. The proposed bandpass filter was a combined inductor and two capacitors including an intertwined spiral inductor, an interdigital capacitor and a spiral capacitor. By equivalent circuit, the series and parallel capacitor can be integrated into one capacitor. Because of this, the proposed filter can be analyzed by first-order inductor and capacitor (LC) BPF as the simulation and measurement result. The performance of the proposed filter was better than other works because the losses were reduced using a GaAs substrate. Especially, the insertion loss was greatly reduced. The manufactured filter using IPD technology made it difficult to know if the filter was fabricated well or not because of its small scale. Through SEM images, we can check that there is no problem in the process.

Table 1 indicates that the proposed BPF has a more compact size and a lower number of metal layers, which reduces the cost and fabrication complexity of the device. Further, the proposed BPF

Electronics **2018**, *7*, 195

had a smaller insertion loss and a higher return loss compared with other reported BPFs. Table 1 shows a comparison of the parameters of the developed BPF with several recently reported IPD BPFs. IPD technology can realize the fabrication of small-sized BPF better than LTCC BPF [21]. Monolithic Microwave Integrated Circuit (MMIC) BPF is almost the same as IPD BPFs [10], but in order to have a high cut-off frequency, the overall size of the filter must be fabricated small. When considering the different center frequency IPD and MMIC technology, IPD BPFs can be realized as a small size and fabricated better than MMIC BPF. The BPF of this work had the lowest insertion loss, and the BPF could be easily and simply fabricated compared to other BPFs because we used only two metal layers.

Table 1. Parameter comparison between proposed BPF and reported BPFs. LTCC, low temperature co-fired ceramic.

Ref.	Technology	Cutoff Frequency (GHz)	Insertion Loss (dB)	Return Loss (dB)	Size (mm^2)	Metal Layers
[19]	Silicon IPD	2.45	2.2	30	1.5 × 1.5	3
[20]	Silicon IPD	1.7	2.0	25	1.0 × 0.5	3
[21]	LTCC	2.4	2.4	15	2.6 × 2.6	4
[9]	Aluminum IPD	2.45	2.8	14	2.2 × 1.8	4
[10]	Al2O3 MMIC	5.2	2.5	29.1	0.81 × 0.68	3
[12]	GaAs IPD	1.63	0.4	24.2	0.88 × 1.0	2
[14]	GaAs IPD	2.27	0.8	26.1	0.9 × 0.7	2
This work	GaAs IPD	1.96	0.27	26.8	0.8 × 9.8	2

4. Conclusions

In order to realize high performance BPFs along with miniaturization and integration required in the mobile industry, IPD technology was used in this study. IPD technology can achieve miniaturization and integration. When using the same substrate, it is possible to fabricate more BPFs that are miniaturized by IPD technology compared with BPFs manufactured by different processes on one substrate. This can lower the price of the BPF, which is advantageous for the mobile industry. To fabricate a BPF with good performance, the number of metal layers is increased. In this study, we used the space of the interdigital capacitor and spiral capacitor inside the circular intertwined spiral inductor by using only two metal layers. By using the GaAs substrate, it is possible to obtain a sharp band pass skirt selectivity with a small insertion loss and a large return loss. By reducing the number of metal layers, the structure is simplified, and the number of manufacturing steps can be reduced to lower the manufacturing cost. In the future, the mobile industry is expected to demand compact BPFs with a better performance. Thus, it is necessary to continue developing IPD technology to meet the future demands of the mobile industry.

Author Contributions: Conceptualization, K.-H.L.; Formal analysis, K.-H.L.; Investigation, E.-S.K.; Methodology, K.-H.L.; Resources, J.-G.L.; Supervision, N.-Y.K.; Writing—original draft, K.-H.L.; Writing—review & editing, J.-G.L. and N.-Y.K.

Funding: This research received no external funding.

Acknowledgments: This research was supported by the basic research project conducted with support from the Korea Research Foundation (NRF) and the government (Ministry of Education) in 2018 (No. 2018R1A6A1A03025242), as well as a 2018 research grant from Kwangwoon University.

Conflicts of Interest: The authors declare no conflict of interest.

References

1. Yeung, L.K.; Wu, K.L. A Compact Second-Order LTCC Bandpass Filter With Two Finite Transmission Zeros. *IEEE Trans. Microw. Theory Technol.* **2003**, *51*, 337–341. [CrossRef]
2. Zhang, S.; Rao, J.Y.; Hong, J.S.; Liu, F.L. A Novel Dual-Band Controllable Bandpass Filter Based on Fan-Shaped Substrate Integrated Waveguide. *IEEE Microw. Wirel. Compon. Lett.* **2018**, *28*, 308–310. [CrossRef]

3. Xue, Q.; Jin, J.Y. Bandpass Filters Designed by Transmission Zero Resonator Pairs with Proximity Coupling. *IEEE Trans. Microw. Theory Technol.* **2017**, *65*, 4103–4110. [CrossRef]
4. Wu, M.C.; Chung, S.J. A Small SiP Module Using LTCC 3D Circuitry for Dual Band WLAN 802.11a/b/g Front-End Solution. In Proceedings of the IEEE Topical Meeting on Silicon Monolithic Integrated Circuits in RF Systems, San Diego, CA, USA, 18–20 January 2006; pp. 174–177. [CrossRef]
5. Ko, Y.J.; Park, J.Y.; Ryu, J.H.; Lee, K.H.; Bu, J.U. A Miniaturized LTCC Multi-layered Front-end Modulefor Dual Band WLAN (802.1 la/b/g) Applications. In Proceedings of the 2004 IEEE MTT-S International Microwave Symposium Digest, Fort Worth, TX, USA, 6–11 June 2004; pp. 563–566. [CrossRef]
6. Kundu, A.; Mellen, N. Miniaturized Multilayer Bandpass Filter with Multiple Transmission Zeros. *IEEE MTT-S Int. Microw. Symp. Dig.* **2006**, 760–763. [CrossRef]
7. Swaminathan, M.; Bavisi, A.; Yun, W.; Sundaram, V.; Govind, V.; Monajemi, P. Design and Fabrication of Integrated RF Modules in Liquid Crystalline Polymer (LCP) Substrates. In Proceedings of the Conference of IEEE Industrial Electronics Society, Raleigh, NC, USA, 6–10 November 2005; pp. 2346–2351. [CrossRef]
8. Sarkar, S.; Palazzari, V.; Wang, G.; Papageorgiou, N.; Thompson, D.; Lee, J.H.; Pinel, S.; Tentzeris, M.M.; Papapolymerou, J.; Laskar, J. RF and mm-wave SOP Module Platform using LCP and RF MEMS Technologies. In Proceedings of the Conference IEEE MTT-S International Microwave Symposium Digest, Fort Worth, TX, USA, 6–11 November 2004; pp. 567–570. [CrossRef]
9. Yook, J.M.; Kim, K.M.; Kwon, Y.S. Suspended Spiral Inductor and Band-Pass Filter on Thick Anodized Aluminum Oxide. *IEEE Microw. Wirel. Compon. Lett.* **2009**, *19*, 620–622. [CrossRef]
10. Chia, S.W.; Chiu, H.C.; Lin, Y.F. Microwave band-pass filter and passive devices using copper metal process on Al2O3 substrate. *Microw. J.* **2008**, *19*, 620–622.
11. Wang, C.; Lee, W.S.; Zhang, F.; Kim, N.Y. A novel method for the fabrication of integrated passive devices on SI-GaAs substrate. *Int. J. Adv. Manuf. Technol.* **2011**, *52*, 1011–1018. [CrossRef]
12. Kim, E.S.; Kim, N.Y. Micro-Fabricated Resonator Based on Inscribing a Meandered-Line Coupling Capacitor in an Air-Bridged Circular Spiral Inductor. *Micromachines* **2018**, *9*, 294. [CrossRef]
13. Li, Y.; Wang, C.; Kim, N.Y. A high performance compact Wilkinson power divider using GaAs-based optimized integrated passive device fabrication process for LTE application. *Solid-State Electron.* **2014**, *103*, 147–153. [CrossRef]
14. Chuluunbaatar, Z.; Adhikari, K.K.; Wang, C.; Kim, N.Y. Micro-fabricated bandpass filter using intertwined spiral inductor and interdigital capacitor. *Electron. Lett.* **2014**, *50*, 1296–1297. [CrossRef]
15. Li, Y.; Wang, C.; Kim, N.Y. Design of Very Compact Bandpass Filters Based on Differential Transformers. *IEEE Micrw. Wirel. Compon. Lett.* **2015**, *25*, 439–441. [CrossRef]
16. Mohan, S.S.; Hershenson, M.D.M.; Boyd, S.P.; Lee, T.H. Simple Accurate Expressions for Planar Spiral Inductances. *IEEE J Solid State Circuits* **1999**, *34*, 1419–1422. [CrossRef]
17. Bryan, H.E. Printed inductors and capacitors. *Tele-Tech Electron. Ind.* **1955**, *14*, 68.
18. Ong, K.G.; Grimes, C.A. A resonant printed-circuit sensor for remote query monitoring of environmental parameters. *Smart Mater. Struct.* **2000**, *9*, 421–428. [CrossRef]
19. Liu, K.; Frye, R.; Emigh, R. Band-Pass-Filter with Balun Function from IPD Technology. In Proceedings of the IEEE Electronic Components and Technology Conference, Lake Buena Vista, FL, USA, 27–30 May 2008; pp. 718–723. [CrossRef]
20. Li, N.; Li, X.Z.; Xing, M.J.; Chen, Q.; Yang, X.D. Design of Super Compact Bandpass Filter Using Silicon-Based Integrated Passive Device Technology. In Proceedings of the IEEE International Conference on Electronic Packaging Technology (ICEPT), Harbin, China, 16–19 August 2017; pp. 1069–1072. [CrossRef]
21. Zhang, X.Y.; Dai, X.; Kao, H.L.; Wei, B.H.; Cai, Z.Y.; Xue, Q. Compact LTCC Bandpass Filter With Wide Stopband Using Discriminating Coupling. *IEEE Trans. Microw. Theory Technol.* **2014**, *4*, 656–663. [CrossRef]

MDPI

St. Alban-Anlage 66

4052 Basel

Switzerland

Tel. +41 61 683 77 34

Fax +41 61 302 89 18

www.mdpi.com

Electronics Editorial Office

E-mail: electronics@mdpi.com

www.mdpi.com/journal/electronics

www.ingramcontent.com/pod-product-compliance
Lightning Source LLC
Chambersburg PA
CBHW051917210326
41597CB00033B/6173